TRANSHUMANISM

TRANSHUMANISM

A Grimoire *of* Alchemical Agendas

Dr. Joseph P. Farrell

&

Dr. Scott D. de Hart

FERAL HOUSE

*"'Why is Love called a Magus?' asks Ficino in the commentary
on the **Symposium**. 'Because all the force of Magic consists in Love.
The work of Magic is a certain drawing of one thing to another
by naural similtude. The parts of this world, like the members of one
animal, depend all on one Love, and are connected together
by natural communion.... From this community of relationship is
born the communal Love: from which Love is born the common
drawing together: and this is the true Magic.'"*

Ficino, *Commentarium in Convivium Platonis de amore*,
oratio VI, cap. 10, p. 1348, cited in Frances A. Yates,
Giordano Bruno and the Hermetic Tradition, p. 127.

*"...the known is a tiny island floating on a vast and very strange sea.
"Let us sow the seeds of doubt. Let us take Francis Bacon's advice and
refrain from rushing to impose a pattern on the world. Let us wait
with Keats at our shoulder for a deeper pattern to emerge.
"Science is **not** certain. It is a myth like any other, representing what
people in the deepest parts of themselves want to believe"*

Mark Booth, *The Secret History of the World
As Laid Down by the Secret Societies*, p. 405.

Transhumanism:
A Grimoire of Alchemical Altars and the Agenda
for the Apocalyptic Transformation of Man

© 2011 by Joseph P. Farrell and Scott D. de Hart

Λ Feral House book
ISBN 978-1-93623-944-3

Feral House
1240 W. Sims Way Suite 124
Port Townsend WA 98368
www.FeralHouse.com
Book design by Jacob Covey

10 9 8 7 6 5 4 3

For

KENNER UND LIEBHABER

TABLE OF CONTENTS

Epigraph v

Dedication vii

PART ONE

Towers and Topology: The Tower of Babel Moment, 3
the Fall of Man, and the Revelation of an Agenda

Introduction: The Disconcerting Images and 5
 Agendas of Alchemy

 A. Alchemo-chimerical Man, Alchemo-Vegetable Man, Alchemosexual 7
  Man: Definitions and Preliminary Observations

 B. "Alchemosexuality" as a Metaphysical First Principle 11

 C. The Term "Alchemosexuality" and the Constellation of Concepts 13
  Embraced in it

 D. The Final Alchemo-Eschatology 15

1. The "Tower of Babel Moment of History": 19
 The Primordial Unity and High Knowledge of Man,
 and How it was Dealt With

 A. The Biblical Version of the Tower of Babel Moment
  1. The Tower of Babel Story Itself 20
  2. And the Fall 21

 B. The Mesopotamian Version: Enmerkar and the Lord of Aratta 22

 C. The Mayan Popol Vuh
  1. The Original Differentiation 25
  2. The Primordial Masculine-Androgyny of Man, Mankind's Original 27
   High Knowledge, and the Fall as Fragmentation

 D. The Platonic Version 29

 E. The Vedic View of the Topological Metaphor and the Fall of Man
  1. The Tree of Life in the Vedas 32

2. The Rig Vedas and the Origin of Sacrifice: 34
    A Metaphor Literally Practiced
3. The Trees of Life and Knowledge in Yahwism   39
F. The Catalogue of Concepts Associated with the Tower of Babel Moment   41

2. THE "TOPOLOGICAL METAPHOR" OF THE MEDIUM AND ITS REVOLUTIONARY   47
INVERSION: THE "FIRST EVENT," THE FOUR-STAGED DESCENT OF MAN, AND
THE THREE GREAT YAHWISMS
A. The Metaphor and the First Event,
    or Primordial Differentiation
    1. The Metaphor and Some of Its Cultural Expressions   48
        a. In Hinduism0
           (1) The Triune Vishnu0   49
           (2) The *Bhagavad Gita*: the Knower and the Field   52
        b. In Egypt
           (1) An Egyptologist Examines the Akhenaton Monotheist   54
               Revolution
           (2) An Esotericist Examines the Traditional Egyptian Cosmology   54
        c. In Mayan Culture   57
        d. In Neoplatonic Tradition   59
        e. The High Esoteric Tradition of the Metaphor:
           The Hermetica and the Image of Androgyny
           (1) God, Space, and Kosmos   60
           (2) Androgyny in the Hermetica   63
        f. Summary of the Metaphor as Examined, and Its   64
           Methodological Implications
B. The Descent of Man
    1. The Universe as the Body of God: Makanthropos, Entanglement,   65
        and the *Bhagavad Gita*
    2. Man as Microcosm   66
    3. Mineral, Vegetable, and Animal Man   66
C. The Esoteric Tradition of the Primordial Unity and   66
    Its Symbol in Androgynous Man
    1. The Primordial "Androgyny" and the   67
        Primary Differentiation
    2. Its Implications   69
    3. The Inverted Implications: The Three Great Yahwisms and the
        Struggle Against the *Prisca Theologia*
        a. The Inversion of the Topological Metaphor to a Technique   70
           of Social Engineering and Construction Via Conflict
        b. Monotheism and the Resulting Social Dualism: 73
           The Convert-Enemy Paradigm of Social Interaction
        c. Nihilism as the Distinguishing Characteristic of Yahwism   76
        d. The Binary Logic of Yahwism Versus the Triadic Logic of the   77
           Metaphor and the Alchemical Eschatological Necessity

3. The Alchemical Agenda of the Apocalypse: 83
   Conclusions to Part One

PART TWO   87
The New Frankensteins:
The Alchemical Ascent to the Animal and the Vegetable:
The Transgenic Transformation of Man

4. Old Homunculi and New Frankensteins: Genetics, Chimeras, and the   89
   Creation of "Alchemanimal" Man
   A. A Brief Overview of The Hermetic Basis of Modern Physics   89
      1. Hermeticism in Copernicus and Kepler   90
      2. In Newton   91
      3. In Leibniz
         a. Leibniz's Characteristica Universalis and the Quest for a   94
            Universal Formal Language
   B. The Alchemical Basis of Modern Genetic Engineering
      1. The Promethean Alchemist's Ambition: The Creation of Life in the   95
         Homunculus
         a. The Dream of reanimation and Virtual Immortality   95
         b. The Androgyne at the End of the Age: The Alchemical   97
            Apocalypse and Final Transformation of Matter
         c. Paracelsus   99
      2. Paracelsus on the "Techniques" of Engineering the Homunculus   101
   C. Conclusions thus Far   102
   D. Chimeras: Alchemanimal Man   103

5. Frankenfoods for the "Alchemo-vegetable" Frankenstein: 113
   the Seedless Seeds of the Androgyne's Food
   A. The Alchemical Background: the Rockefellers and Francis Bacon   114
   B. Esoteric Eugenics, Banksters, Seedless Seeds, and
      Alchemovegetable Man
      1. A Babylonian Theme Revisited: Too Many People(?)   118
      2. The Vipers of Venice Reiterate the Theme: Carrying Capacity   119
      3. The Banksters Adopt the Babylonian Theme   119
      4. The Rockefellers, the "Food Weapon," and the Alchemical Seedless Seeds
         a. A World War, and the "Peace Studies Group"   121
         b. An Esoteric Connection?   122
         c. The "Food Weapon" and Other Techniques of Alchemical   122
            Social Engineering
         d. Argentina: The Alchemical Laboratory for the Seedless Seeds
            of Alchemy
            (1) The Historical Background: The Beginning: 124
                The Rockefellers, Nazis, and Perón

(2) The Rockefellers and Argentina — 125

C. Genetically Modified Crops and The Patent Weapon:
Patented Plants, Pigs, and, Maybe People?

  1. Secret Meetings and "Substantial Equivalence" — 127

6. THE TRANSHUMANIST TECHNO-ANDROGYNY AND THE — 135
ALCHEMOMINERAL MAN

A. DARPA's GRINs: A Brief Review of the Background of Transhumanist
Technologies

  1. The Keys to Creation — 136

    a. A Form of Magical Reversal of the — 137
Tower of Babel Moment

    b. Man the Microcosm becomes Man the Macrocosm — 138

    c. Downloading and Uploading Memories, and Direct — 138
Modification of Consciousness and Behavior

    d. The Group Consciousness? Or Roddenberry's "Borg"? — 140

  2. DARPA's Quest for the Transhumanist Supersoldier — 141

    a. Longevity — 142

    b. DARPA's and the Defense Science Office's Mission Briefs — 143

    c. Back to Longevity — 143

B. The Scenarios of the Transhumanist Apocalypse Theater

  1. The Three Scenarios — 145

  2. The Assumptions of the Scenarios of the Transhumanist Apocalypse — 145
Theater

C. The Possibilities and Dangers of a Breakaway Civilization: — 147
"Enhanced" vs. "Normal" Humans

7. RECAPITULATION: CONCLUSIONS THUS FAR — 153

PART THREE — 157
THE ANDROGYNOUS GOD OF ALCHEMY AND THE ALCHEMOSEXUAL ASCENT

8. "AQUINAS," ALCHEMOSEXUALITY AND THE ANDROGYNOUS GOD: — 159
TRANSMUTATION, TRANSUBSTANTIATION, AND TRANSITIONS

A. Aquinas' Theology-Stopping Vision — 160

B. The Alchemosexual Vision of the *Aurora Consurgens* and Its Implications

  1. Attributions and Authenticity of the *Aurora Consurgens* — 162

  2. The Style of the *Aurora Consurgens,* and a Problem — 163

  3. The Seventh Parable and the Alchemosexual Androgyny of — 165
God and Man

    a. The Twofold Movement — 171

    b. The Implications — 172

C. Transubstantiation and the Topological Metaphor — 173

An Introductory Interlude: 177
Frankenstein's Alchemical Fiction:
Shelley Unbound and The Picture of Oscar Wilde,
An Introduction to Chapter 9

9. Monsters, Pictures, and Pits: Percy Bysshe Shelley, Oscar Wilde,
Dante Alighieri and the Tower of Babel Moment
A. Monsters and Myths   179
B. The Monster within the Man   186
C. Alchemy and *Frankenstein's* Meaning   193
D. Shelley's *Epipsychidion*, Or, The Soul-less Creature   199
E. The Picture of Oscar Wilde   202
F. Autobiography of an Alchemist   204
G. Dante's Pit and Climb out of Hell   213

10. The Esoteric Androgyny: Secret Societies and the Hidden Tradition   221
of Alchemosexual Man
A. Alchemosexuality in Masonic Initiation
   1. The First Degree: Entered Apprentice   221
   2. The Second Degree: Fellow Craft and the Gradual Revelation of the   224
      Primordial Alchemosexuality
   3. The Master Mason and the Full Revelation of the Primordial   227
      Alchemosexuality in the Context of Geometry and the
      Topological Metaphor
B. The Rosicrucians   230
C. Joachim of Fiore, the Hidden Androgynous God, and the Antinomian   235
   Kingdom of the Spirit: The Bridge to Alchemical Messianic Expectation

11. The Androgynous Apocalypse: Hermaphroditism, Uranians, Shamans,   243
and Geneticists
A. The Nineteenth Century Explosion of Hermaphroditism   246
   1. "Gonadism"   248
   2. Androgyny: Pathology? or Third Sex?
      a. Embryonic Androgyny   249
B. The Uranianian-Darwinian Guess
   1. Edward Carpenter, Uranianism,   251
      and the Topological Metaphor
   2. Erasmus Darwin: "The Temple of Nature," and the Evolutionary   258
      Algorithms of Differentiation
C. The Behavioir Genetics Confirmation, or Falsification?   262
D. Shamans, Drugs, DNA, and Androgyny   265
E. Objections: Religion and Yahwism Again   270

12. Epilogue is Prologue: Microcosm and Medium: The Anthropic   279
Cosmological Principle in Physics

TRANSHUMANISM

I.

Towers and Topology:

The Tower of Babel Moment, the Fall of Man,

and the Revelation of an Agenda

"Everything has been organized by the monad, because it contains everything potentially: for even if they are not yet actual, nevertheless the monad holds seminally the principles which are within all numbers..."

—Iamblichos the Neoplatonist, The Theology of Arithmetic
Trans. Robin Waterfield (Kairos Press, 1988), p. 35.

⚜ Introduction ⚜

The Disconcerting Images and

Agendas of Alchemy

*"You are now in this Degree permitted to extend your
researches into the more hidden paths of nature and science."*
From the Initiation of the Fellow Craft[1]

IF WE WERE TO TELL YOU that there is an alchemical agenda that tran-
scends religions, secret societies and spans the millennia, you would probably
- and rightly - at least entertain the suspicion that we had taken momentary
leave of our sanity. If we were to tell you, further, that this alchemical agenda
spans virtually every discipline that you can think of - from biology to history,
physics, topology, art, music - even, as we shall see in the main text, literary
criticism - you would probably entertain that idea more seriously, for that,
indeed, is what we are going to tell you in these pages, for superintending all
the alchemical images and their implied agendas that we survey here, there is
one standing out above them all, that both compels the agendas, and simultaneously reveals some of them as forms of false alchemy; the image is that of
primordial simplicity, or androgyny, or "Nothingness," or physical medium,
or aether, or "ocean of quantum flux," or Grand Architect of the Universe.
The image goes by many names, depending on the fashion of the age, and the
particular agenda emphasized, but it is, nonetheless, the same image.

This means that modern man is in a predicament, for he is about to be
sacrificed, either upon an apocalyptic altar of alchemical science, or, if one is
to believe the "Three Great Yahwisms" - Judaism, Christianity, and Islam -
slaughtered by a righteous God come back to restore justice to the world by
an unparalleled bloodletting, prior to mankind's final transformation - if one
is to believe a certain strand of Christian fundamentalist eschatology - into
the very same sorts of alchemical creations as proffered by the transhuman-

ist science they excoriate. Either way, the transhumanist gospels of Science or the revelations, prophets, and ministers of Yahwism are saying the same thing: the New Age is here; prepare to die as part of your process of alchemical transformation.

For those caught in the middle, neither worshipping the unrestrained power lust of modern science, nor the bloodthirsty "God" of the "revealed monotheisms," this is, indeed, a predicament.

In this book, we propose to examine the first pole of this sacrificial dialectic, the "scientific" one, reserving our comments on the bloody eschatology of the Three Great Yahwisms for a future work and sequel to this, though we shall, of course, treat it briefly here. Nonetheless, we *are* concerned here with altars and an alchemical, transhumanist apocalypse, for in this case, the altars are not only in churches, but the altar in preparation is the earth, the sacrifice is mankind, and the alchemy is...well... alchemy, for there is nothing terribly modern about the goals and agendas being discussed and advanced in "modern" science at all. In the Introduction to our previous book, *The Grid of the Gods,* we wrote;

> Modern science is but a technique of the imagination to bring into reality the operations of the magical intellect and the mythologies of the ancients, with consistent and predictable regularity. This implies, therefore, that the magical intellect encountered so often in ancient texts, myths, and monuments is, in fact, the product of a decayed science, but a science nonetheless. Much of modern physics may be viewed as but Hermetic metaphysics with "topological" equations,[2] and by a similar process of examination, much of modern genetics may be viewed as but the myths of Sumer, Babylon, and even the Mayans, given flesh by the techniques of genetic engineering.[3]

Though we both noticed this odd coincidence of modern science and ancient myths, each of us came to the writing of this book by very different, and yet in many respects, by very parallel routes. Indeed, for both of us, a heavy atmosphere of synchronicity hovers over the observations and experiences that brought us here, to this book. For each of us, the process began when we were both at Oxford – though at different times – pursuing our PhD's in theology, and we each kept what we were noticing carefully to ourselves, in hidden thoughts written in notebooks of observations, carefully held away from public view or hidden even more carefully away on the tablets of our minds. There, like Percy Shelley's[4] tormented monster-creator, Victor Frankenstein, we began to "notice" things in the writings of Medieval schoolmen, alchemists, theologians, and novelists:

> These thoughts supported my spirits, while I pursued my undertaking with unremitting ardour. My cheek had grown pale with study, and my person had become emaciated with confinement. Sometimes, on the very brink of certainty, I failed; yet still I clung to the hope which the next day or the next hour might realize. One secret which I alone possessed was the hope to which I had dedicated myself, and the moon gazed on my midnight labours, while, with unrelaxed and breathless eagerness, I pursued nature to her hiding places. Who shall conceive the horrors of my secret toil …[5]

We too were after an answer regarding how the ancients understood Nature, life, death, and the creation or re-creation of life., and quite naturally so, for as students of theology and philosophy, our journey was to be a focus of our doctoral research as well as lead us down paths that, until now, were shared only between the two of us in private conversations.

A. Alchemo-chimerical Man, Alchemo-vegetable Man, Alchemosexual Man: Definitions and Preliminary Observations

Why was it, we wondered, that the basic ancient myths - excepting those offshoots of Yahwism - were based on the idea of mankind's descent from a Primordial Androgyny, through the Mineral, through the Vegetable, and finally into the Animal Kingdoms? How could the ancient myths even *speak* of a "Mineral Man" or "Vegetable Man?" These images are disconcerting, even nonsensical, but the most disconcerting thing about them, as we discovered, was that they also reappeared in the snapshots of modern science that every so often make a column filler-article in a newspaper, or make their way around internet sites.

Why was it, we wondered, that ancient esoteric lore could speak of the lowest level of mankind's descent, that of the Animal Kingdom, populating its mythological world with chimerical hybrid creatures, half animal, and half human, and modern science could speak of the same things - even calling them "manimals" - as a *goal to be sought?*

Why was it, we wondered, that ascending from this to the Vegetable Kingdom - which paradoxically was viewed in the ancient lore as a higher kingdom than the animal - why was it that the ancient images could speak of strange "androgynies," of "fusions" between the plant and human, and in modern times, why is it that genetically-modified plants sprouting their own kind of alchemical "seedless seeds", or plants modified with splices of human DNA, were touted, once again, as a *goal to be sought,* as a good thing that would be a boon to mankind? We wondered: Was it possible that the whole agenda of modern science was from top to bottom an alchemical agenda for the complete transformation of mankind? Was this genetically modified food for the alchemically modified man?

Ascending from there to the Mineral Man, again we wondered: why were so many within the "transhumanist" movement seeming to speak, through all their modern verbal coinage, nothing but the language of ancient lore and alchemy? Once again, there seemed to be an agenda that was nothing less than a quest for an "androgynous fusion" of man and machine, of man and the mineral.

This brought us to a consideration of the most disconcerting image of them all: androgyny itself, which more often than not stood for a fusion of many sorts of paired principles that seemed at first unrelated to the concept of androgyny in its most basic sense. It stood, in other words, precisely for that fusion of the human and the animal, of the human and the vegetable, of the human and the mineral, as it stood also for the fusion or union of the masculine and the feminine, of the male and the female.

We realized that we needed a whole new vocabulary even to be able to *discuss* the alchemical connections and roots of all these things: "alchemochimerical man" to designate the transformation of mankind into a chimerical creature via the techniques of science, "alchemovegetable man" to discuss the fusions of man and plant, again via the techniques of science; "alchemomineral man" to designate the alchemical fusion of man and machine, again, by the techniques of applied science, for in each case, science is functioning as nothing but an extension of alchemy - itself a technique - for the transformation of mankind, which is, of course, a primary goal of alchemy and its modern equivalent, "transhumanism." For us, the term "transhumanism" really serves only to mask what is a very old and explicitly alchemical, apocalyptic agenda. In short, while the *techniques* of "science" may be more refined than those of the "pseudoscience" of alchemy, the goals remained essentially and existentially the same.

That left the most disconcerting image of them all, one found in nearly every culture and religion, even the Jewish and the Christian, though it is very carefully disguised, and the subject carefully avoided. That image was the image of the primordial androgyny, both of God, and of Man.

Why was it, we wondered, that so many of the initiatory rites of secret societies and fraternal organizations within the occidental tradition were overtly, and yet subtly, "alchemosexual" in nature(and patience, we will explain the strange term "alchemosexual" shortly)? One need think only of the allegations of Skull and Bones initiates at Yale stripping naked and wrestling in mud,[6] or, as we shall see in the main text, of the androgynous implications of Masonic initiation rites? Why is it, moreover, that those societies so often restricted their memberships to men?

The mystery deepened the more we looked. As Joseph delved into research and authored a whole series of books on ancient physics and history, again and again, and to his consternation, he bumped into what can only be described as a "primordial masculine androgyny," oftentimes associated with alchemical doctrines, that is to say, a view of the gods (or of God), or even of the prime matter or "soup," that was deeply, explicitly, and simultaneously masculine, androgynous, and alchemical both in its symbolism and in its implications. Oftentimes, this imagery would appear the most explicit precisely in those religions and cultures most strongly condemning alchemy, alchemical or esoteric societies, and alchemosexual rituals or ceremonies. One need only consider the Christian Trinity as such an androgynous image, laden with multiple levels of meanings and potential misunderstandings, and how, by contrast, the Roman Church, for one, has more or less officially and (almost) universally condemned any esoteric alchemical practices or fraternities, especially those in which life appears to imitate the art of the mysterious image. At the minimum, it seemed to us that there was a "disconnect" between the effeminate "Jesus image" of Roman Catholic religious iconography, and the ethic.

However, other religions, adopting a similar imagery, followed the principle of "life imitating art," and thus freely approved of or instituted corresponding rituals, ceremonies, and life practices consistent with the alchemosexual imagery related to the god(s). How then, did one account for the difference? And at what point did it arise? And why? This too, was a mystery: why did some promote such masculine images, then qualify them by theological caveats stating that God was beyond or above such sexual distinctions and was in fact a kind of androgyne in their own version of the "masculine-androgyny;" why then, at the same time, did some religions and cultures insist on the permanence and revelatory character of this image, and therefore establish it as the final character of those masculine-androgynous gods - the alchemosexual images themselves - and *then* go on to condemn the rituals, ceremonies and corresponding practices which are implied in them? Why, in other words, did some accept it, and others reject it, when the symbolism to depict the divine was explicitly "alchemosexual"? Indeed, what led the ancients even to conceive of the oxymoronic idea of a "masculine androgyny" in the first place, and then to regard it as a kind of metaphysical, indeed, even *physical* first principle?

Over several years of such discussion, we began to suspect that there was more going on here than met the eye, and began to explore the possibility that this image may have been a profound cosmological clue, perhaps even a residue of very old doctrines, a legacy coming from High Antiquity, and that the image was therefore not original to the later classical and esoteric systems promoting it. We began to suspect that maybe the imagery had less to do with religion than with metaphors about the underlying physical substrate or *materia prima* itself, with the "God behind God" as it were.

Then, a second mystery was added to compound the first. As our study was theology, early Christian history, doctrine and ritual, we were familiar with a little-known – one is tempted to say, deeply and deliberately buried – early tradition that mankind himself was originally created as a kind of masculine-androgynous "alchemosexual" creature that was capable of some sort of reproduction. As will be seen in the main body of this book, variations of this tradition held that the division of the sexes was a result of the Fall of Man, or accomplished in pre-vision for it. The implication of such a view, again, was disturbing to say the least, for implicit in that idea is the notion that any reproduction by the original "masculine-androgynous" man would by the nature of the case have been "homosexual" or "asexual" in nature.

As we dug, we soon realized that this concept also had a far older provenance than just the Christian, or for that matter, just the Jewish, traditions. It was also a traditional teaching of the deepest and oldest mystery schools and was a principal component of very old esoteric and occult doctrines, from Egypt to China to Meso-America. The clear impression that *these* gave was, once again, that the idea came from remotest antiquity. Thus, the mystery deepened: when, and why, was this tradition obfuscated, and when, and why, did persecution of any manifestation of alchemosexuality - metaphorical or otherwise - emerge? Coupled with this problem was another: the emergence of bloody, and oftentimes, human sacrifice.

As we continued to research down the long avenues of esoteric tradition and ancient hermetic texts, we encountered yet other clues, clues indissolubly connected at every turn to this very ancient image of the "primordial masculine alchemosexuality", and indeed, we have eventually come to view these clues as the three most closely held secrets at the core of such fraternities, secret societies, and mystery schools, dating back into the mists of "High

Antiquity." The first of these secrets was, as already mentioned, that in these ancient cosmological systems, the primordial symbol both of God and of man is "alchemosexual" in nature, not as a matter of faith or revelation, but as a matter of a kind of "formally explicit knowledge" about the physics of the medium itself. However, coupled always to this in the ancient view were, curiously, two other "secrets" or "mysteries" namely:

1) That God, or at least, "Someone" or "Something" or(depending upon the particular interpretation in view) "Nothing" exists, not as a matter of faith, but as a matter of a kind of "formally explicit" knowledge; and,
2) that personal immortality also exists, not as a matter of faith, but again, as a kind of "formally explicit" knowledge.

We were not, we tentatively concluded, looking at anything theological or metaphysical in the conventional senses of the terms, nor were we looking at anything merely sexual, but rather, at something that encompassed sexual, spiritual, and metaphysical, and even biology and physics, components into one confusing if not disconcerting whole.

With respect to the first of these "three secrets," the curious and ambiguous wording is necessary due to the extraordinary nature of the statements made in those ancient texts and cosmologies. One was dealing with systems that defied conventional analysis or pigeon-holing into a "theistic" or "atheistic" box, but rather, as will be seen in the main text, with systems that could fit into both at the same time. The images and cosmologies in view were, in other words, a kind of "acid drip" on all conventional techniques of dialectical philosophical analysis; they were components of a deep cosmic, and anthropological, *ritual*. They defied convention and tradition precisely because they claimed to be the oldest conventions and traditions. All others were reductions to one particular *subset* of implications.

B. "Alchemosexuality" as a Metaphysical First Principle

For us, the question then became: why should all three of these things be so persistently, consistently, and alchemically linked, from the *Vedas* of India, the *I Ching* of China, to the *neters* of Egypt and even to the *Popol Vuh* of the Maya in Meso-America and the emanations of the *Hermetica*, of Plato, and of the Neoplatonists? The widespread diffusion of the "alchemosexual" symbolism of God or man defied any conventional *diffusionist* model; it was an argument that one had, perforce, to be looking at a symbolic legacy coming down from

High Antiquity, or at the minimum, at a kind of Jungian archetype in the human mind itself, or both.

As we researched further, more questions surfaced. Why, for example, was there an emergence of this imagery in the poetry and prose of the otherwise reserved and staid English Victorian era? Repeatedly, throughout 19th century English literature, one encounters images so disturbing to "normal" sensibilities that the authors were condemned for their work. Two of the nineteenth century "men of letters" we will consider (though the list of such authors could be far larger and require a book unto itself) wrote gothic novels with alchemy, apocalyptic revolutionary visions, and forbidden love as a lurking shadow behind the main characters: namely, Percy Bysshe Shelley and Oscar Wilde. Most biographers have accurately noted that, both with Shelley and Wilde their personal lives and the art which was inspired by it was anything but "normal" by the standards of their day. Even today some would find Shelley and Wilde's philosophy of "love" far beyond the acceptable norm, and it is precisely at this point that our research attempts to indicate correlations between alchemy and an ideal for the higher man, a god-like man. Unlike the State and the Church, we are neither condemning certain literature as immoral nor are we passing judgment on the lifestyle of the authors of such controversial literature. We do seek, however, to understand *why* they were moved to write novels with alchemical agendas and imagery which would unnerve the fragile conscience of their readers. Was the life or even the love that was sought by their protagonists or even antagonists (not to mention the authors themselves) "higher" or was it dangerously subversive to the future of mankind?

In our search for an answer to that question we discovered that even within the Christian patristic and Gnostic traditions,[7] there were clear allusions to the same idea, allusions taken over almost whole cloth from very ancient, and very *non*-Christian, sources. Again, we encountered the imagery from the Mayans to the ancient Vedic Indians. The question for us, then, is *why* is this there, and *how might one rationalize it?* Here the key, oddly enough, lies in ancient cosmologies, in the *physics* and a profound "topological metaphor" that we have explored in previous books.[8]

The presence of this metaphor is in itself perplexing, for it cannot be gainsaid that it is not only both *ancient* and fairly universal, having every appearance of having come from High Antiquity, but also that the metaphor itself stresses the fact that this "primordial alchemosexual androgyny" *is* primordial, i.e., **it is regarded as a kind of metaphysical "first principle" by almost all who employed it.** And again, the question is, *why is it there in the first place?* In this book, we attempt to argue a speculative answer to these questions.

C. The Term "Alchemosexuality" and the Constellation of Concepts Embraced in It

But this fact, to our mind, was perplexing for a very different reason, for it was clear to us that we needed to coin a whole new vocabulary to *talk* about it, without falling into the trap of advocating any sort of position toward it. Again, our need was to *rationalize* the thought process, not *justify* it. We were not alone in feeling this need, for as will be seen in the main text, the nineteenth century "Uranian" scholars felt the same need too. For us, however, the need arises not merely from the need to avoid contemporary terms which perforce concentrates attention only and merely upon a physical or sexual phenomenon and thus avoids the spiritual, aesthetic, and social implications implied by the primordial metaphor in all its fulness, but it also arises out of the fact that the metaphor finds expression in the esoteric tradition, and even in secret societies.

In short, we needed a term that would designate simultaneously the masculine-androgyny, its association with esoteric doctrines, secret fraternal societies, and with very ancient *and* very modern cosmological views and with their whole presentation of a "ladder of descent " from heaven, a descent that implies the process of "reverse engineering" and a re-ascent up the same ladder.

Edward Carpenter, in his now classic early twentieth century study of this whole problem, *The Intermediate Sex: A Study of Some Transitional Types of Men and Women,* cited a statement of Xavier Mayne, in which Mayne clearly intuited a deep connection between this "alchemosexual-masculine androgyny" and the fraternal tradition of secret societies:

> I realised that I had always been a member of that hidden brotherhood and Sub-Sex, or Super-Sex. In wonder too I informed myself of *its deep instinctive freemasonries* - even to organised ones - in every social class, every land, and every civilisation.[9]

What Mayne intuited only vaguely by his reference to "its deep instinctive freemasonries," our investigations - after encountering the imagery in *The Grid of The Gods* - soon revealed was a major component of ancient esoteric tradition, and a thinly veiled alchemosexual ritual in some fraternal societies.

During the course of our mutual conversations and researches over the years on alchemy and related subjects, we came to another, equally disconcerting realization: alchemy was identical in almost all respects with the basic fundamentals of Christian sacramentalism:

- Both insisted that matter could be transformed into the vehicle for oneness with the divine;
- Both aimed therefore not only at the transformation of matter, but eventually at the apocalyptic transformation of mankind himself from "base metal" into "immortal gold";
- Both insisted that a basic "recipe" be followed, employing the proper matter for this transformation, performed at certain specific times, and according to certain specific formulae of words and the proper moral intentions.

Where the two *differed* was that alchemy insisted that there was no need for special revelations, churches, or priesthoods to accomplish all of this. Rather, it was a goal pursued through the millennia by "technological" means. No Church other than that of "Nature" and "Nature's God" were necessary. The one was a ritual of revaltion; the other a ritual of technique and technology.

It is when one considers the full implications of these points that the disconcerting realization begins to dawn, for there is implicit in these propositions the idea that there is a hidden agenda concerning knowledge, power, technology and the final transformation of man according to its own "alchemosexual" principles. Indeed, so disconcerting are the details of these principles and agendas that we do not, in this book, even come close to encompassing all the minutiae; the knowledge is too powerful, and the danger too high.

That said, we needed a term that thus could encompass the following things in addition to the two factors noted previously:

3) the primordial, metaphysical, spiritual, and even basic *physics* nature of the metaphor, both as it was applied to God or the Gods, or to the primordial *materia prima*, and, for that matter, to primordial mankind;

4) the *persistent masculinity* of the metaphor, as distinct from those readings of the esoteric and mythological tradition that emphasize the primordial *femininity*, an interpretation we find somewhat suspect for reasons argued in the main text;

5) the association of the metaphor with esoteric traditions and fraternities, both in the sense of conventional secret societies, and more broadly, as "sub-cultures";

6) the association of the metaphor with *immortality* and the apocalyptic and hidden agenda of the alchemosexual transmutation of man; and finally,

7) its obvious sexual implications.

We have thus coined the term "alchemosexuality" to denote and encompass this entire constellation of concepts, and their deep connections to esoteric and alchemical traditions, societies, rituals, and *agendas*.

D. The Final Alchemo-Eschatology

As we pondered all these images, one final thing - the most deeply disturbing thing of them all - became clear: the final political and alchemical transformation of mankind himself seemed to be the hidden goal of so many political movements, and even a hidden implication of the various systems of "end-time" speculations of various religions. While we do not propose to examine these exhaustively in this book, we do propose to lay the groundwork for a future examination of them, by looking briefly at the hints of the emergence of these political agendas in the Middle Ages through the Renaissance, leaving their hidden influences upon modern religion and politics for a future book.

In conclusion, we would remind the reader of our real purpose: we would have the reader clearly understand that this book does *not* propose nor advocate the pursuit or practice of any of these alchemical goals, much less practice of a "lifestyle," or anything of the sort. We are concerned *solely and exclusively* with the exposition and exploration of the disconcerting images both of God and of Man found throughout history and in some very unexpected places, from literature, mythology and religious iconography, to secret societies, and with the influences of that imagery. We are concerned solely with understanding its possible roots and implications for the field of alternative research, and for the possible activities of hidden elites through the millennia, in a secret, fraternal continuity. Consequently, we attempt herein not to *justify* the perplexing images or metaphors, but rather, simply to *rationalize* them, to *reconstruct a **possible** thought process* that led the ancients to formulate them in the first place, and modern science to *revivify* that pursuit. We are thus also concerned to rationalize the basis for the influence of that image - oftentimes scarcely or little appreciated - over the human imagination.

That said, we do not mince words in our analysis or our critique, for we believe that it is high time that people confront the implications of these images and their social implications directly, and deal with them with genuine compassion and tolerance, and not with the outworn response of denunciation or persecution founded on "standard answers" that bear little relation to the deep roots of this tradition. Accordingly, our rationalization of the deep roots of this alchemical imagery, and of the possible elites employing or manipulating it, is of course speculative, and highly so. Nonetheless, we do believe that these roots, stemming from a profound metaphor of a deep physics, and the application of the analogical method by the ancients to understand it, is at least close to the mental processes that led to the formulation of these images and cosmologies.

Therefore, we do not claim our analysis is complete, only that it is highlighting obvious though overlooked things, largely because people do not wish to face them. We do not assume that we will convince anyone because we are not seeking to convince, but to only to explore and rationalize a complex and curious phenomenon, a veritable galaxy of constellations of concepts, images and traditions that are all closely associated. We do not presume to write as a theologians or clergymen on behalf of any Church or religion. We are authors challenging models of history and thought, whether religious or secular, in order to provoke deeper thought related to modern research. We do not call people to any faith, but to those of faith or without faith, this book is intended to provoke thought, not a following or belief. But we do hope that we will, on the end of this work, have at least made some people stop and think.

ENDNOTES

1   Citing Robert Boyle, *Works*, Robert Lomas, *Freemasonry and the Birth of Modern Science* (Fair Winds Press, 2003), p. 65

2   See my *The Giza Death Star Destroyed* (Adventures Unlimited Press, 2005), pp. 222-245, and my *The Philosophers' Stone* (Feral House, 2009), pp. 42-48.

3   Joseph P. Farrell with Scott D. de Hart, *The Grid of the Gods* (Adventures Unlimited Press, 2011), p. iii.

4   The authors are aware that Mary Wollstonecraft Shelley was until recently credited as the sole author. Recent evidence from the Frankenstein Notebooks at the Bodleian Library, University of Oxford, has given Percy Bysshe Shelley partial credit based on notations in his own hand to the novel. The authors hold to the opinion that Percy Bysshe Shelley is the *sole* author of the 1818 first edition based on textual and extra textual evidence and that the Mary Shelley authorship is a hoax that was perpetrated by none other than Shelley himself. Cf. *The Man Who Wrote Frankenstein*, John Lauritsen, Pagan Press, 2007.

5   Mary Wollstonecraft Shelley, *Frankenstein: The Original 1818 Text*, ed. D.L. Macdonald and Kathleen Scherf, Second Edition (Broadview Literary Texts: 1999), p. 82.

6   The most comprehensive book on Skull and Bones and similar fraternities at Yale University is Kris Millegan, Ed., *Fleshing Out Skull and Bones: Investigations into America's Most Powerful Secret Society* (Walterville, Oregon: TrineDay, 2003). For the reference to wrestling naked in a mud pile, see Anthony Sutton, *America's Secret Establishment: An Introduction to the Order of Skull and Bones* (Billings, Montana: Liberty House Press, 1986, ISBN 0-937765-02-3], p. 201.

7   We mean "patristic tradition" in the academic sense, not an ecclesiastical one, for the writings cited herein are not always from those acknowledged by either Roman Catholicism, Anglicanism, or Orthodoxy, as saints or doctors of the church and teachers of its doctrine.

8   See Joseph P. Farrell, *The Philosophers' Stone: Alchemy and the Secret Research for Exotic Matter*(Feral House, 2009); Joseph P. Farrell with Scott D. de Hart, *The Grid of the Gods: The Aftermath of the Cosmic War and the Physics of the Pyramid Peoples* (Adventures Unlimited Press, 2011).

9   Xavier Mayne, *Imre: a memorandum* (Naples: R. Rispoli, 1906, pp. 134-135), cited in Edward Carpenter, *The Intermediate Sex: A Study of Some Transitional Types of Men and Women* (London: George Allen & Co., Ltd., 1912), p. 169.

≋ One ≋

THE "TOWER OF BABEL MOMENT OF HISTORY":

THE PRIMORDIAL UNITY AND HIGH KNOWLEDGE OF MAN,

AND HOW IT WAS DEALT WITH

∴

*"If God is a tribal, racial, national, or sectarian archetype, we are the warriors of his cause; but if he is a lord of the universe itself, we than go forth as knowers to whom **all** men are brothers."*
—Joseph Campbell[1]

"Obviously, the ultimate evidence of universality in speech is to be sought in the oldest languages of man, prehistoric languages which preceded, so to speak, the Tower-of-Babel moment in human evolution."
Leonard Bernstein[2]

VIRTUALLY EVERY RELIGION, and most esoteric metaphysical systems, have had something to say about one of the most curious topics in religion: the Tower of Babel Moment of history, or, to put it in more familiar and conventional terms, the Fall of Man. It is a curious, perhaps even a bizarre, image, when one really stops to think about it, because of the constellations of concepts that usually accompany it, and this is true, no matter where one looks.

Due to our cultural matrix, we in the West tend to view the Fall of Man and the Tower of Babel moment as separated events, for that is the way they are presented in the Bible. But we believe that they are, in fact, connected fragments of a single story, perhaps separated out of some editorial agenda.

What is also equally curious, is that virtually all these concepts refer to a point in history, where mankind was engaged in a project, or where his knowledge and/or unity posed some sort of implicit threat to God or the gods.

We will look at various versions of this Tower of Babel, Fall of Man Moment of History, as they are recorded in disparate and discrete traditions, for in doing so, a very interesting catalogue of concepts emerges.

A. The Biblical Version of the Tower of Babel Moment
1. The Tower of Babel Story Itself

While we have written about this topic elsewhere,[3] it is best to review those scattered comments from the perspective of what we are calling "The Tower of Babel Moment," for when viewed synoptically, these various traditions reveal an intriguing list of related concepts, and in doing so, reveal a hidden alchemical agenda, consisting of precise steps, to re-cement a primordial human unity, a unity lost in the fragmentation of the "Tower of Babel Moment." Thus, while this section will be, to some extent, a review of what we have previously written, the review is necessary to establish the catalogue of related concepts and thereby this hidden "alchemical eschatology" that those concepts imply.

The Old Testament version of the Tower of Babel Moment is recorded in Genesis 11:1-9:

> [1] And the whole earth was of one language, and of one speech. [2] And it came to pass, as they journeyed from the east, that they found a plain in the land of Shinar; and they dwelt there. [3] And they said one to another, Go to, let us make brick, and burn them thoroughly. And they had brick for stone, and slime had they for mortar. [4] And they said, God to, let us build us a city, and a tower, whose top may reach unto heaven; and let us make us a name, lest we be scattered abroad upon the face of the whole earth. [5] And the LORD came down to see the city and the tower, which the children of men builded. [6] And the LORD said, Behold, the people is one, and they have all one language; and this they begin to do, and now nothing will be restrained from them, which they have imagined to do.
>
> [7] Go to, let us go down, and there confound their language, that they may not understand one another's speech. [8] So the LORD scattered them abroad from thence upon the face of the earth; and they left off to build the city. [9] Therefore is the name of it called Babel; because the LORD did there confound the language of all the earth:

and from thence did the LORD scatter them abroad upon the face of the earth.

What is unusual about this story it that is lacks the type of moral sanctions usually given in the Old Testament for Yahweh's - symbolized by the translation "LORD" in the Authorized Version - actions.

Such sanctions are missing here; rather, what one is left with is that somehow, whatever it is that mankind is doing, it requires action to fragment his unity and stop the project. A glance at the concepts implied in the story will be helpful in highlighting this dynamic:

1) Mankind is viewed in a state of *unity* which is exemplified by a common speech, and this unity is *fragmented* by the action of Yahweh, by creating a multiplicity of language;

2) The "unified speech" of mankind might also imply a unified language of science, i.e., a highly unified scientific worldview wherein the major sciences - physics, biology, genetics, and so on - are all viewed and understood with a completeness and unity our current science lacks;

3) The idea of language also subtly implies the idea of *sound* and this might be connected to the Tower; confounding the languages thus breaks not only mankind's unity but perhaps also breaks or impedes the power of the Tower;

4) The story implies that a unified mankind is a *threat* in some sense, and that this threat is tied to the Tower itself;

5) Thus, the fragmentation of man via confounding of his language removes both the threat and its power.

2. And the Fall

While the Tower of Babel story is separated in the Old Testament from the story of the Fall of Man, nonetheless the same broad list of conceptual relations obtains between the two, and this suggests that they are meant to be at least conceived as related stories, if not, perhaps, two different tellings of the same event, for the points enumerated above are also subtly in play in the biblical story of the Fall of Man in Genesis chapter three, where again there is a coupling of knowledge and the loss of the unity of mankind, this time through the imposition of death, or the fragmentation of soul and body, and as we shall encounter much later, some rabbinical and patristic commentators even argue that the division of the sexes from a "primordial masculine androgyny" was also connected to the Fall.

B. The Mesopotamian Version:
Enmerkar and the Lord of Aratta

The Mesopotamian version of the Tower of Babel Moment of history is given in the epic *Enmerkar and the Lord of Aratta.* Here the story emerges as a contest between two cities, Uruk and Aratta, with Enmerkar, priest-king of Uruk demanding, from the city of Aratta, workers, gold, and, as we shall see, an unusual kind of silver to complete a tower. Throughout the story, however, the androgynous goddess Innana hovers in the background, along with the ever-present and mischievous Enki, who was the god that helped engineer mankind into existence in other Mesopotamian texts.[4]

For example, at the beginning of the epic we read that Innana carried "the great mountain" in Uruk in her heart.[5] Determining to coerce Arrata to supply the workers and materials needed, Enki and Innana issue a veiled threat:

> May Enki not have to
> > curse Aratta
> > and the settlements,
> may he not have to
> > destroy it too
> > like places he had
> > (at other times) destroyed.
> Innana has set out after it,
> has scre(amed at it,)
> > (has ro)ared at it,
> may she (not) have to
> > dwrown it (too)
> > with a flood wave
> > (like) the flood waves
> > (with which she drowns).[6]

Shortly after this scarcely veiled threat is issued to Aratta, we find the first reference in the text to an implied unity of mankind:

> In those days
> > there being no snakes,
> > there being no scorpions,
> there being no hyenas,
> > there being no lions,
> there being no dogs, or wolves,

there being no(thing) fearful
>   or hair-raising,
mankind had no opponents -
in those days
>   in the countries Subartu
>   Hamazi,
bilingual Sumer
>   being the great country
>   of princely office,
the region of Uri
>   being a country
>   in which was
>   what was appropriate,
the country Marduk
>   lying in safe pastures,
(in) the (whole) compass
>   of heaven and earth
>   *the people entrusted (to him)*
could address Enlil,
>   *verily, in but*
>   *a single tongue.*[7]

Here, as in the biblical version, mankind's unity appears to be couched in *linguistic* terms, though how Sumer could be both "bilingual" and yet speak to Enlil "in but a single tongue" is not made clear.

As in the biblical text, it is this socio-linguistic unity of mankind that is broken; the one who breaks it is Enki:

In those days
>   (having) lordly bouts,
>   princely bouts, and royal bouts -
(did) Enki, (having) lordly bouts,
>   princely bouts, and royal bouts -
having lordly bouts fought,
>   having princely bouts fought,
>   and having royal bouts fought,
did Enki, lord of abundance,
>   lord of effective comment,
did the lord of intelligence,
>   the country's clever one,

did the leader of the gods,
did the sagacious
    omen-revealed
lord of Eridu
estrange the tongues
    in their mouths
      as many as were put there.
The tongues of men
    which were one.[8]

Unlike the biblical version, no reason is really given as to why the gods - or Enki acting alone - took the decision to fragment mankind's unity by scrambling his single language into several. However, a hint of the reason might lie in the suggestions of conflict recorded at the beginning of the passage. If this be so, then mankind apparently, again, may have posed some sort of threat to the "gods" in his unified state.

Before departing this story, there is a curious passage that must be mentioned, for it contains the only hint of a *technology* in the story. First, the threat is repeated by Enmerkar's envoy to Aratta, and then a demand for certain materials is made:

May I not have to
    made his city
    fly off from him
    like wild doves from their tree,
may I not have to
    make it fly
    like birds out of their nests,
may I not have to appraise it
    at current market rate (for slaves).
May I not have to
    (scoop up) dust in it
    as in a destroyed city.
May Enki not have to
    curse Aratta
    and the settlements,
may he not have to
    destroy it too,
    like places he has
    (at other times) destroyed.

Inanna has set out after it,
has screamed at it,
    has roared at it.
May she not have to
    drown it too
    with a flood wave
    with which she drowns.
Rather, when it has packed
    gold in its native form
    into leather pouches,
has aligned with it
    purified silver
    in dust form...[9]

Those familiar with the lore of alchemy will recognize this reference to *powdered* metals, in this case, silver "in dust form", the final stage in the confection of the Philosophers' Stone.

Additionally, metals in powdered form are a hallmark of the so-called "ORMEs" or "Orbitally Rearranged Monatomic Elements," first discovered by Arizona farmer David Hudson. The nuclei of such metals apparently exist in a high-spin state, possess peculiar anti-gravity properties, and can only be confected by extremely high heat.[10] The appearance of "purified silver in dust form" thus strongly suggests a high technology in play, and an alchemical one at that.

The *Enmerkar and the Lord of Aratta* epic thus only hints at certain themes made more clear in the biblical version:

1) Mankind existed in a *primordial unity* that was *linguistic* and therefore presumably social in nature;
2) Mankind's unity was *fragmented* by Enki through the confounding of languages, *after* the barest of hints of conflict;
3) Further threats are issued against Aratta in the name of the androgynous goddess Innana and the god Enki;
4) A slight though clear suggestion of an alchemical technology is made in the demands on Aratta to deliver a purified silver in dust form.

C. The Mayan Popol Vuh
1. The Original Differentiation

Turning from the Mesopotamian to the Meso-American, the Mayan *Popol Vuh* records yet another, and one of the most intriguing, versions of the Tower of Babel Moment. But before we can examine that more closely, we must situate it in its context of its account of the primary differentiation itself. In our previous book, *The Grid of the Gods*, we noted that this primary differentiation in Mayan legend takes a specific form, and it is worth recalling in detail what we said about it there:

> Like the Hindu cosmology laid out in stone reliefs at Angkor Wat, The *Popol Vuh* begins in an abyss of mystery, an abyss laid out in eloquent and elegantly simple words and imagery whose power is made even more manifest by their poetic simplicity:

> This is the account, here it is:
> Now it still ripples, now it still murmurs, ripples, it still sighs, still hums, *and it is empty under the sky.*
> *Here follow the first words, the first eloquence.*
> *There is not yet one person, one animal, bird, fish, crab, tree, rock, hollow, canyon, meadow, forest. Only the sky alone is there;* the face of the earth is not clear. *Only the sea along is pooled under all the sky; there is nothing whatever gathered together. It is at rest;* not a single thing stirs. It is held back, kept at rest under the sky.
> *Whatever there is that might be is simply not there: only murmurs, ripples, in the dark,* in the night. Only the Maker, Modeler alone, Sovereign Plumed Serpent, the Bearers, Begetters are in the water, a glittering light. They are there, they are enclosed in quetzal feathers, in blue-green.
> Thus the name, "Plumed Serpent." They are great knowers, great thinkers in their very being.
> And of course there is the sky, and there is also the Heart of Sky. This is the name of the god, as it is spoken.
> And then came his word, he came here to the Sovereign Plumed Serpent, here in the blackness, in the early dawn.... Thunderbolt Hurricane comes first, the second is Newborn Thunderbolt, and the third is Sudden Thunderbolt.
> So there were three of them...[11]

> Note ... that the *topological metaphor* of a primordial trinity is preserved. Everything begins as an emptiness "under the sky" and there is not yet any differentiation within it: "there is not yet one person, one animal" and so on. There is only an empty sky, and pooled water at rest beneath it. The only thing existing is Sovereign Plumed

Serpent and a mysterious reference to "Bearers" and "Begetters in the water" who are described as "great knowers, great thinkers in their very being," who are later found, just like Vishnu, to be manifestations of Sovereign Plumed Serpent.

The *Popol Vuh* is telling us, in other words... (that) there is a primordial "nothing", Sovereign Plumed Serpent, and then there is a primordial "trinity," of endless indistinct "sky" and below it a "sea", *and the implied common surface between the two.* Nothing else whatsoever, at this juncture, exists, except a faint "murmuring" and "rippling" in the night, implying somehow that *sound, frequency, vibration* give rise to all the fecund distinctions and variety to follow.

Indeed, at the very beginning, the *Popol Vuh* informs us that "This is the beginning of the Ancient Word, here in this place called Quiché. Here we shall inscribe, we shall *implant* the Ancient Word, *the potential and source for everything done...*in the nation of the Quiché people."[12]

2. The Primordial Masculine-Androgyny of Man, Mankind's Original High Knowledge, and the Fall as Fragmentation

It is within this context that the *Popol Vuh* sets the creation, and subsequent fragmentation, of mankind:

And these are the names of *our first mother-fathers. They were simply made and modeled,* it is said; they had no mother and no father. We have named the men by themselves. No woman gave birth to them, nor were they begotten by the builder, sculptor, Bearer, Begetter. *By sacrifice alone, by genius alone they were made, they were modeled by the Maker, Modeler, Bearer, Begetter, Sovereign Plumed Serpent.* And when they came to fruition, they came out human:
They talked and they made words.
They looked and they listened.
They walked, they worked.
They were good people, handsome, with looks of the male kind. Thoughts came into existence, and they gazed; their vision came all at once. *Perfectly they saw, perfectly they knew everything under the sky, whenever they looked.* The moment they turned around and looked around in the sky, on the earth, everything was seen without any obstruction. They didn't have to walk around before they could see what was under the sky: they just stayed where they were.

As they looked, their knowledge became intense. Their sight passed through trees, through rocks, through lakes, through seas, through mountains, through plains.....

And then they were asked by the builder and mason:

"What do you know about your being? Don't you look, don't you listen? Isn't your speech good, and your walk? So you must look, to see out under the sky. Don't you see the mountain-plain clearly? So try it," they were told.

And then *they saw everything under the sky perfectly*. After that, they thanked the Maker, Modeler:

"Truly now,
double thanks, triple thanks
that we've been formed, we've been given
our mouths, our faces,
we speak, we listen,
we wonder, we move,
our knowledge is good, we've understood
what is far and near,
and we've seen what is great and small
under the sky, on the earth.
Thanks to you we've been formed,
we've come to be made and modeled,
our grandmother, our grandfather,"

they said when they gave thanks for having been made and modeled. *They understood everything perfectly, they sighted the four sides, the four corners in the sky, on the earth, and this didn't sound good to the builder and sculptor:*

"What our works and designs have said is no good:

'We have understood everything, great and small,' they say." And so the Bearer, Begetter took back their knowledge:

"What should we do with them now? Their vision should at least reach nearby, they should see at least a small part of the face of the earth, but what they're saying isn't good. Aren't they merely 'works' and 'designs' in their very names? Yet *they'll become as great as gods, unless they procreate, proliferate at the sowing, the dawning, unless they increase.*"

"*Let it be this way: now we'll take them apart just a little,* that's what we need. What we've found out isn't good. *Their deeds would become equal to ours, just because their knowledge reaches so far....*

And such was the loss of the means of understanding, along with the means of knowing everything, by the four humans. The root was implanted.

...

*And **then** their wives and women came into being.*[13]

Once again, we note the themes of the unity and knowledge of mankind are linked, and that in that condition, mankind poses some sort of threat to the gods. As a result, we find again the themes of the loss of knowledge and the fragmentation of mankind are linked in a kind of fall.

But there are a number of crucial details here.

1) The unity of mankind is conceived to be in masculine-androgynous terms, i.e., as an original male-female sexual unity, and this unity is tied, somehow, to the "perfect" knowledge that mankind has, a point which is stressed over and over again in the passage;

2) This unity and knowledge in turn constitute some sort of threat to the gods;

3) The gods take the decision to curb mankind's original unity in the now familiar pattern, by fragmenting mankind, only in this case, the fragmentation is, predictably, the division of the original androgyny into the sexes. Once this is accomplished, mankind loses his knowledge. The implication is that mankind's original androgyny might have been tied either to immortality or longevity, and the division of the sexes results in the loss of that longevity or immortality, and consequently, in a loss of knowledge.

It is also to be noted that mankind is created by some act or action conceived to be a sacrifice. All these themes find their earliest expression, once again an ocean and half a world away, in Plato's *Symposium*, and in the ancient culture of the Vedas.

D. The Platonic Version

The Platonic version of this primordial androgyny and Fall is more difficult to piece together, simply because the whole doctrine is scattered throughout the various dialogues of Plato. For our purposes, we shall concentrate on only one of his dialogues here, the *Symposium* or *Banquet*, placing its contents within the wider Platonic system. Oddly, however, Plato suggests almost exactly the same thing as is stated in the Mayan *Popol Vuh*, namely, the division of the sexes had to be accomplished, so far as the gods were concerned, in order to render mankind "more feeble":

> You ought first to know the nature of man, and the adventures he has gone through; for his nature was anciently far different from that which it is at present. First, then, human beings were formerly not divided into two sexes, male and female; there was also a third, common to both the others, the name of which remains, though the sex itself has disappeared. The androgynous sex, both in appearance and in name, was common both to male and female; its name along remains, which labours under a reproach.
>
> At the period to which I refer, the form of every human being was round, the back and the sides being circularly joined, and each had four arms and as amany legs; two faces fixed upon a round neck, exactly like each other; one head between the two faces; four ears, and two organs of generation; and everything else as from such proportions it is easy to conjecture. Man walked upright as now, in whatever direction he pleased; and when he wished to go fast he made use of all his eight limbs, and proceeded in a rapid motion by rolling circularly round, - like tumblers, who, with their legs in the air, rumble round and round. We account for the production of three sexes by supposing that, at the beginning, the male was produced from the Sun, the female from the Earth; and *that sex which participated in both sexes, from the Moon, by reason of the androgynous nature of the Moon....*
>
> *They were strong also, and had aspiring thoughts. They it was who levied war against the Gods; and what Homer writes concerning Ephialtus and Otus, that they sought to ascend to heaven and dethrone the Gods, in reality relates to this primitive people. Jupiter and the other Gods debated what was to be done in this emergency. For neither could they prevail upon themselves to destroy them, as they had the Giants, with thunder, so that the race should be abolished; for in that case they would be deprived of the honours of the sacrifices which they were in the custom of receiving from them; nor could they permit a continuance of their insolence and impiety.* Jupiter, with some difficulty having devised a scheme, at length spoke. 'I think,' said he, 'I have contrived a method by which we may, by rendering the human race more feeble, quell the insolence which they exercise, without proceeding to their utter destruction. I will cut each of them in

half; and so they will at once be weaker and more useful on account of their numbers....'[14]

Again, we have a similar catalogue of concepts as in the Mayan *Popol Vuh*:

1) There is a primordial human androgyny, only in Plato's case, at the beginning of the cited passage, this apparently exists along side of the other two sexes;
2) This primordial adrogyny, as with the Mayans, constitutes some sort of *threat* to the gods, as they were "strong" and had "aspiring thoughts" even to the point of wanting to wage war against them and dethrone them, thus implying that this creature had some sort of knowledge by dint of its androgyny that was a threat to the gods;
3) The gods deliberate on what to do, and, as in the Mayan account, decide that abolishing the race altogether - as they had done with the Giants or Titans - was out of the question since mankind would no longer be able to offer them *sacrifice*, i.e., it is implied that mankind's relationship to the gods is once again one of *debt*;
4) The decision is taken, at the suggestion of Jupiter(Zeus) that this primordial androgynous man be "cut in half," i.e., that the sexes should be divided.

With this decision made, the dialogue continues, seeking to explain human sexual behavior according to the Platonic doctrine of recollection (ἀναμνησις) of a former state of existence and its fall from the higher realm of the ideals:

Immediately after this division, as each desired to possess the other half of himself, these divided people threw their arms around and embraced each other, seeking to grow together; and from this resolution to do nothing without the other half, they died of hunger and weakness: when one half died and the other was left alive, that which was thus left sought the other and folded it to its bosom; whether that half were an entire women (for we now call it a woman) or a man; and thus they perished. But Jupiter, pitying them, thought of another contrivance, and placed the parts of generation before. Since formerly when these parts were exposed they produced their kind not by the assistance of each other, but like grasshoppers, by engendering upon the earth. In this manner is generation now produced, by the union of male and female' so that from the embrace of a man and woman the race is propagated, from those of the same sex no such consequence ensures.

... Every one of us is thus the half of what may be properly termed a man, and... is the imperfect portion of an entire whole, perpetually necessitated to seek the half belonging to him. Those who are a section of what was formerly one man and woman, are lovers of the female sex.... Those women who are a section of what in its unity contained two women, are not much attracted by the male sex, but have their inclinations principally engaged by their own.... Those who are a section of what in the beginning was entirely male seek the society of males; and before they arrive at manhood, such being portions of what was masculine, are delighted with the intercourse and familiarity of men.....

The cause of this desire is, that according to our original nature, we were once entire.[15]

As we shall discover in section three of this book, there is an odd sort of modern scientific corroboration of this idea of primordial androgyny and "recollection." For now, however, we must turn to the Vedic culture.

E. The Vedic View of the Topological Metaphor and the Fall of Man
1. The Tree of Life in the Vedas

In the *Upanishads*, there occurs an intriguing passage on "the tree of life":

The Tree of Eternity has its roots above
And its branches on earth below.
Its pure root is Brahman the immortal
From whom all the worlds draw their life, and whom
None can transcend. For this Self is supreme!

The cosmos comes fort from Brahman and moves
In him. With his power it reverberates,
Like thunder crashing in the sky. Those who realize him
Pass beyond the sway of death.[16]

The image of the tree and Brahman occurs again in the fifteenth chapter of the *Bagavad Gita*, where Sri Krishna is speaking:

There is a fig tree
In ancient story,
The giant Aswattha,
The everlasting,
Rooted in heaven,
Its branches earthward:
Each of its leaves
Is a song of the Vedas,
And he who knows it
Knows all the Vedas.

Downward and upward
Its branches bending
Are fed by the gunas,
The buds it puts forth
Are the things of the senses,
Roots it has also
Reaching downward
Into this world,
The roots of man's action.

What its form it,
Its end and beginning,
Its very nature,
Can never be known here.

Therefore, a man should contemplate Brahman until he has sharp-ened the axe of his non-attachment. With this axe, he must cut through the firmly-rooted Aswattha tree.... Let him take refuge in that Primal Being, from whom all this seeming activity streams forth forever.[17]

Commenting on this image of the tree, Paramahansa Yogananda states the following:

"Trees" symbolize the bodies of all living things - plants, animals, man - possessing their own distinct type of roots, trunks, and branches with their life-sustaining circulatory and nervous systems. Of all living forms, only man's body with its unique cerebrospinal centers has the potential of expressing fully God's cosmic consciousness. The sa-cred Ashvattha tree... therefore symbolizes the human body, supreme among all other forms of live.

Man's physical-astral-causal body is like an upturned tree, with roots in the hair and brain, and in astral rays from the thousand-petaled lotus, and in causal thought emanations which are nourished by cosmic consciousness. The trunk of the tree of life in man is the physical-astral-causal spine. The branches of this tree are the physical nervous system, the astral *nadis* (channels or rays of life force), and thought emanations of the magnetic causal body. *The hair, cranial nerves, medulla, cerebral astral rays, and causal thought emanations are antennae that draw from the ether life force and cosmic consciousness. Thus is man nourished not only by physical food, but by God's cosmic energy and His underlying cosmic consciousness.*[18]

In other words, Yogananda is suggesting that human DNA is itself the "tree of life" and the "tree of knowledge," acting in a manner somewhat analogous to a radio receiver, transducing or "tuning into" a particular sub-set of the information in the field of the cosmic consciousness, or God, as the unique personhood of the individual. This tree of life is also eternal,[19] and thus, is also a tree of immortality, transducing through the body and the mind,[20] as it were, the immortality of the Cosmic consciousness to the individual person, allowing the latter, with the illumination of knowledge, to attain immortality in the ultimate "alchemical" transformation.

Here, as in so many other ancient philosophies and mythologies, mankind is an original primordial androgyny, whose experience of sensuality actually causes the fall, the loss of knowledge, and the division into the sexes.[21] In some versions of this Fall "from androgyny" into "sexual division," it is Brahma who destroys the knowledge that androgyny brings (and here let us understand that androgyny is not only a symbol of the fusion of sexes but of *other* polarities: motion and rest, being and becoming, and so on), and who replaces the implied idea of *communion in consciousness and love* with the idea of *sacrifice.* And with sacrifice, we are in the presence of yet another one of those "disconcerting images."

2. The Rig Veda and the Origin of Sacrifice: A Metaphor Literally Practiced

Throughout our survey of ancient texts containing the topological metaphor in *Grid of the Gods*, and in particular when we encountered the Mayans and Aztecs, the notion of bloody and indeed human sacrifice was tied to the metaphor, implying that by this brutal and barbaric practice immortality could somehow be attained and the gods appeased. As we have seen, the practice was *not* the original practice, as least as far as the Aztecs were concerned. So whence, and why, did it originate, and when? More importantly, how can it be rationalized as a "development" of the metaphor, or can it be? If so, is it a valid, or twisted, development and adaptation?

To answer these questions, we must turn to some of the oldest texts in the world that explicitly mention sacrifice, and that do so explicitly in connection to the topological metaphor itself: the *Rig Vedas*.

In his absolutely crucial and magisterial study - *Meditations through the Rg Veda* - Antonio de Nicolás cites the Vedic hymn, the *Purusa Sukta*, the "hymn of man." We italicize and boldface the portions of this hymn that will concern us in our analysis and speculative reconstruction for why the practice of bloody - and human - sacrifices emerged:

1. Thousand headed is Man,
    With thousand eyes and feet,
    He envelopes the whole earth
    And goes beyond it by ten fingers.

2. *Man indeed is all that was and is,*
    And whatever may come in the future,
    He is the master of immortality,
    Of all that rises through nourishment.

3. Such is his power and greatness,
    Yet man is still greater than these:
    Of him all the worlds are only one-fourth,
    Three-fourths are immortal in Heaven.

4. With three-fourths of Himself, Man rose,
    The other fourth was born here.
    From here on all sides he moved
    Toward the living and the non-living.

5. *From him was Viraj born,*
    And Man from Viraj.
    When born he overpassed the earth,
    Both in the west and in the east.

6. *When with Man as their offering,*

The Gods performs the sacrifice,
Spring was the oil they took
Autumn was the offering and summer the fuel.

7. *That sacrifice, balmed on the straw,*
   Was Man, born in the beginning;
   With him did the gods sacrifice,
   And so did the Sadhyas and the Rsis.

8. *From that cosmic sacrifice,*
   Drops of oil were collected,
   Beasts of the wing were born,
   And animals wild and tame.

9. *From that original sacrifice,*
   The hymns and the chants were born,
   The meters were born from it,
   And from it prose was born.

10. From that horses were given birth,
    And cattle with two rows of teeth.
    Cows were born from that,
    And from that were born goats and sheep.

11. **When they dismembered Man,**
    Into how many parts did they separate him?
    What was his mouth, what his arms,
    What did they call his thighs and feet?

    ...

16. By sacrifice the gods sacrificed the sacrifice.
    Those were the original and earliest acts.
    These powers (of the sacrifice) reach heaven,
    Where the Sadhyas and the gods are.[22]

Summarizing the emphasized points reveals an interesting picture and set of relationships:

1) God, or the primordial medium, is viewed in effect as a "grand

man" or "cosmic man," that is, as a "makanthropos" (μακανθροπος) (stanzas 2 and 5);

2) This "makanthropos" is "the master of immortality"(stanza 2), in other words, immortality and the medium are intimately connected for reasons as yet to be explored and understood;

3) Man is himself the offering and sacrifice, and from the context it emerges that the "Man" referred to here is the "cosmic man" or "makanthropos"(stanzas 6-8);

4) It being the "cosmic Man" that is sacrificed, the sacrifice itself is "cosmic"(stanza 8) and thus the sacrifice has the power to "reach heaven" (stanza 16), implying the power to *affect* the heavens, that is to say, the divine or the medium, in some fashion;

5) The sacrifice of this cosmic Man consisted of his *dismemberment* (stanza 11), yet another disconcerting image in what has now become a very long list of disconcerting images.

De Nicolás produces more references as to the importance of sacrifice in the *Rig Vedic* system:

> ...(The) Rg Vedic(sic) seers place us face to face *with what is primary to man: the first act of man, the Sacrifice*: "With sacrifice the gods begot the first one, and it became the first Act of mankind (1.164.50) In this way, the One came to be spoken of as many: "They call it Indra, Mitra, Varuna, Agni and Garutmat the heavenly bird (the Sun)." (1.164.46) And it is in the sacrifice that the past and the future coincide: "Future ones are also ancient, some say, and those part are also present." (1.164.19)[23]

In other words, the sacrifice referred to in the *Purusa Sukta*, and all the implications enumerated above, is the first activity, and one may infer, the primary activity and function of man.

But all of this, it should be noted, is stated in the context of the original topological metaphor of the physical medium, and thus, this imposes certain interpretive limitations on how to understand these sacrificial images and references. De Nicolás thus notes that the images of sacrifice are themselves a metaphor of this original topological metaphor, a metaphor of a metaphor:

> Decapitation, dismemberment, and Sacrifice are also identified in (the *Rig Veda*) 1.52.10; 2.11.2; 2.20.6; 4.19.3....
> ....

It is in the "midst of the (three) homes of Agni (that) the breathing swift-moving, living, restless enduring One" is found, and that the "mortal has a common origin with the immortal."... The different images of perception, either as confused or non-differentiated in *Vrtra*, or differentiated as *Purusa, Prajapati*, Indra, *Soma*, etc., all end up in the Sacrifice - through decapitation, dismemberment, interaction or as the sensorium synthesis.[24]

Further on, De Nicolás clarifies what all these very difficult references to non-differentiation and differentiation mean:

> This returning to the original infinite space ... is no longer the return to inaction, but rather the result of action, an action leading to that illumined instant-moment of light... where the "Father and the Mother meet," where ... Heaven and Earth unite in a common nest, since, after all, "the mortal and the immortal have a common origin."[25]

In other words, *sacrifice is a metaphor of the original "primary scission" or first differentiation which resulted in the rise of differentiation itself.*

Viewed in this context, "dismemberment" and "decapitation" are poetic, if grizzly, codes for "differentiation" and thus sacrifice - at least in the cosmic and original sense - refer simply to that primary and first differentiation that leads, within the topological metaphor, to the rise of all other diversities.

But it is well-known that *actual* sacrifice was indeed practiced in ancient times in Vedic India. So how might one rationalize its rise? The actual practice of sacrifice, as elaborated by De Nicolás, was thus viewed as a kind of "reverse" engineering designed to effect the unions of various diversities, heaven and earth, father and mother, and so on.

We are therefore bold to suggest that the rationalization of the *practice* was rather simple. As the original act of differentiation within the metaphor was perceived as an act of Love - for where there is no differentiation there cannot be any Love - then in the absence of love, the metaphor *came to be understood literally*, and the actual practice of sacrifice became perceived as the means whereby to analogically reproduce the processes of the medium and its differentiations itself, and thereby to *affect* or to "traumatize and shock" the physical medium. We are bold to suggest that this also can only be rationalized by positing the existence of those of evil intention, who viewed the practice simply as a means of acquiring power through this practice of "analogical magic."

We are, in short, once again in the presence of the agenda of communion and union through actual acts of love, versus that of "communion" and "union" through a technique of sacrifice, whose true purposes and motivations are altogether different. The goal or agenda remains the same - the alchemical recreation of a higher alchemosexual union between "opposites"- but the methods of getting there are entirely different.

To put it differently, and much more bluntly, within the context of the alchemosexual-topological metaphor, *sacrifice is the alchemosexual act itself,* a physical though figurative "dimemberment" and sewing of the "seed," the masculine element, into the "ground," the feminine element. It is a bloody sacrifice, because indeed the male seed in most cases contains a small issue of blood.

This means something else, equally and profoundly disconcerting to one's normal sensibilities but nevertheless is a component of this ancient metaphor, for it means that the institution of *actual* bloody sacrifices was a perverted imitation, an act that could never issue love, for it was based neither in differentiation, nor resulted in it, but rather, annihilated it; it was a literal analogical dismemberment designed to traumatize the medium, versus a figurative analogical "dismemberment" designed to deeply penetrate the medium - the ground, the earth, the feminine, in alchemical texts, the "putrefying dung", and even the *actual* female - and to subtly influence it. This means too, that in some traditions both homosexual and heterosexual acts were conceived to be alchemosexual acts; but bloody sacrifice could *never* be. In the latter instance, we have the possible rationalization for the rise of actual bloody sacrifices in Vedic India, and in the former the possible rationalization for the rise of homosexual tantric sex magic in Tibet. It also rationalizes, at least partially, why so many cultures viewed the androgynous manifestation of it as "higher" than the heterosexual, for to the ancient mind, to the mystery schools, the fraternities, and the high tradition of esotericism, it was a *closer* analogue to the androgynous physical medium differentiating itself than the heterosexual act was perceived to be. But more of this in section three. For the present, we must deal with an interloper on the scene.

3. The Trees of Life and Knowledge in Yahwism

It is worth pausing to compare the images of the Tree of Life and Knowlededge - the *themes* of life and knowledge - in the ancient philosophical tradition versus the biblical-Yahwist traditions (Judaism, Christianity, and Islam). The famous scholar of mythology, Joseph Campbell, stated the fundamental difference between the older *philosophia perrenis* and the new Yahwism this way:

The principle of mythic dissociation, by which God and his world, immortality and mortality, are set apart in the Bible is expressed in a dissociation of the Tree of Knowledge from the Tree of Immortal Life. The latter has become inaccessible to man through a deliberate act of God, whereas in other mythologies, both of Europe and of the Orient, the Tree of Knowledge is itself the Tree of Immortal Life, and, moreover, still accessible to man.[26]

In the older more ancient view, it is "not only the individual, but all things" that are "epiphanies"[27] of the primordial reality, of the physical medium, the *materia prima* or God or Cosmic Consciousness, to employ Yogananda's term, from which they are descended.

But with Yahwism, which, like Brahmanism, substituted bloody sacrifice for communion, a tremendous inversion occurs:

According to our Holy Bible, on the other hand, God and his world are not to be identified with each other. God, as Creator, made the world, but is not in any sense the world itself or any object within it, as A is not in any sense B. There can therefore be no question, in either Jewish, Christian, or Islamic orthodoxy, of seeking God and finding God either in the world or in oneself. That is the way of the repudiated natural religions of the remainder of mankind: the foolish sages of the Orient and wicked priests of Sumer and Akkad, Babylon, Egypt, Canaan, and the rest...[28]

The result of this is a complete change of mental outlook in the cultures influenced by Yahwism and its inversions of traditional mythological symbols:

The type of scholarship characteristic of both the synagogue and the mosque, therefore, where the meticulous search for the last grain of meaning in scripture is honored above all science, never carried the Greeks away. In the great Levantine traditions such scholasticism is paramount and stands opposed to the science of the Greeks: for if the phenomenal world studied by science is but a function of the will of God, and God's will is subject to change, what good can there possibly be in the study of nature? The whole knowledge of the first world principle, namely the will of God, has been by the mercy of God made known to man in the book that he has furnished. Ergo: read, read, read, bury your nose in its blessed pages, and let pagans kiss their fingers at the moon.[29]

In other words, once a claim to a special revelation is made, whether by a particular god claiming to be the original undifferentiated God, or whether by a "prophet" or elite, that special revelation will replace and supplant the primordial philosophy, and its system will become paramount in its intellectualized world, seeking to defend the system above all else, either winning converts, or labeling all others as enemies of or infidels to that system. We shall have more to say on this in the next chapter.

F. The Catalogue of Concepts Associated with the Tower of Babel Moment

Putting all this together, one emerges with a rather interesting catalogue of concepts that are usually associated with the "Tower of Babel-Fall of Man" moment of history:

1) The Tower of Babel-Fall of Man moment is always tied to some notion of *the fragmentation* of mankind,
   a) whether it be the division of sexes, as explicitly noted in the *Popol Vuh*, or subtly implied in Genesis 2-3; or,
   b) whether it be by the fragmentation of man's being by death and the disunion of soul and body; or,
   c) whether it be by some sort of social fragmentation, as in the case of the confounding of man's languages in the biblical version of the Tower of Babel Moment;
2) This fragmentation led to a loss of knowledge on mankind's part, implying that whatever *prior unity* that mankind had somehow contributed to some sort of state of advanced knowledge;
   a) in the case of the biblical story of the confounding of languages and the linguistic and social fragmentation, this makes particular sense, since such a process would slow down and seriously impeded the advance of human knowledge and social institutions;
   b) in the case of the fragmentation of mankind through death, and the subsequent decline of human life-spans, this too makes reasonable sense, for greatly shortening human life spans means that the overall progress of knowledge will be much slower since the entire sum of human knowledge has to be recycled and passed down to a new human generation;
   c) in the case of the Mayan *Popol Vuh*, the effect of the division of the sexes on human knowledge is not immediately clear,

though, as we shall see in a later section, speculative rationalizations can be made;

3) What is equally curious is that many religious and philosophical traditions conceive of the Tower of Babel Moment of History precisely *as a **fall**,* that is, they conceive and describe a "spiritual" condition in terms of a *spatio-temporal movement.*[30] This is one of those "obvious things" that becomes a profoundly important physics clue, for as we shall discover, the ancients did *not* understand this movement in the metaphorical terms of modern academics. On the contrary, they *meant* it;

4) Also tied to the idea of the loss of *knowledge* that occurred with the fragmentations of mankind and the various *types* of traditions employed to describe that fragmentation - the male-female fragmentation, the soul-body fragmentation, the linguistic and social fragmentation - is the idea that mankind suffered also a considerable decline in *power,* an unusual motif for religions to be centered upon;

5) What is one of the most *unusual* and indeed, at first, arresting and disturbing, facets of various traditions of the Tower of Babel Moment, is that whatever action is undertaken by the gods to cause the fragmentation of mankind is undertaken *not* for "spiritual" reasons - the standard "unrighteous man offending the righteous gods" theme - but rather, that the gods undertake it because of an explicitly stated or implied threat that mankind, his knowledge, and/or his activity, poses to them. This, oddly enough, is particularly the case in the biblical version of the Tower of Babel Moment;

6) It is thus our belief that because one finds the same constellation of concepts split among various traditions (or even split within the *same* tradition, as in the biblical accounts of the Fall and Tower of Babel Moment as two separate events), that we are looking at fragments of *one* story, of which each tradition preserves some distinctive elements.

It is important to realize what all this means, in terms of those techniques and technologies, for as the Tower of Babel Moment resulted in increasing diversification, inevitably, any re-ascent will involve the conceptual notion of the reunion of distinctions going back up the ladder of descent, of man and animal, of man and plant, of man and the mineral, or machine, and, at the highest level of male and female, of mind and matter, and ultimately issue in a quest for the final, eschatological and alchemical transformation of human consciousness. *In other words, one has but to look at the list of concepts involved in the Tower of Babel Moment, and one may perceive the alchemical, transformation agenda:*

(1) *the recovery of lost knowledge,*
(2) *therewith the recovery of the god-threatening power that went with it,*
(3) *the reunion of elements perceived, rightly or wrongly, to have been at one time united.* As we shall discover in the coming pages, this implies an agenda, one to reunite mankind with each aspect of the ladder of his topological descent in the High Esoteric Tradition.

However, before we can understand that engineered technological ascent as a hidden alchemical agenda, we must have a closer look at the "original event" that got everything started.

Endnotes

1   Joseph Campbell, *The Hero With a Thousand Faces* (Novato, California: New World Library: 2008), p. 138.

2   Leonard Bernstein, *The Unanswered Question: Six Talks at Harvard* (Harvard University Press, 1976), p. 12.

3   Joseph P. Farrell and Scott D. de Hart, *The Grid of the Gods: The Aftermath of the Cosmic War and the Physics of the Pyramid Peoples* (Adventures Unlimited Press: 2011); Joseph P. Farrell, *The Giza Death Star Destroyed: The Ancient War for Future Science* (Adventures Unlimited Press, 2005); *The Cosmic War: Interplanetary Warfare, Modern Physics, and Ancient Texts* (Adventures Unlimited Press, 2006.

4   For Enki's role in engineering mankind into existence, see Joseph P. Farrell, *The Cosmic War: Interplanetary Warfare, Modern Physics, and Ancient Texts* (Adventures Unlimited Press, 2007), pp. 140-149, and *Genes, Giants, Monsters, and Men* (Feral House, 2011), pp. 138-155..

5   Thorkild Jakobsen, trans. and ed., *The Harps that Once....: Sumerian Poetry In Translation* (Yale University Press: 1987), p. 280. The association of the goddess Innana with "the great mountain" of Uruk recalls the formula "Mountains " PLanets " Gods " Pyramids" that was first discussed in Joseph's *The Cosmic War* (pp. 74-83; 232-233; 239-240293-294).

6   Jakobsen, op. cit., pp. 287-288.

7   Ibid., pp. 289-290, emphasis added.

8   Ibid., p. 290, emphasis added. The identification of Enki as the god responsible for dividing mankind's linguistic unity bears its own relation to the biblical version, if one recalls David Rohl's suggestion that Yahweh might *be* Enki. See Joseph P. Farrell, *The Cosmic War*, pp. 301-303.

9   Ibid., pp. 292-293, emphasis added.

10 See Joseph P. Farrell, *The Giza Death Star Destroyed* (Adventures Unlimited Press: 2005), pp. 151-174, and *The Philosophers' Stone: Alchemy and the Secret Research for Exotic Matter* (Feral House, 2009), pp. 85-119.

11 Dennis Tedlock, transl., *Popol Vuh: The Definitive Edition of the Mayan Book of the Dawn of Life and the Glories of Gods and Kings* (New York: Simon and Schuster, 1996), pp. 64-65, emphasis added.

12 Ibid., p. 63, emphasis added.

13 Ibid.,, pp. 146-148, emphases added.

14 Plato, *The Banquet*, trans. Percy Bysshe Shelley (Provincetown, MA: Pagan Press, 2001), pp. 47-48, emphasis added. We have utilized the great poet Shelley's translation as it is one of the few that does not gloss the translation and its implications for androgyny.

15 Ibid., pp. 49-51.

16 Katha Upanishad, 3:1-2, *Upanishads* , Trans. Eknath Easwaran (Blue Mountain Center of Meditation, 1987), p. 95.

17 Swami Prabhavananda and Christopher Isherwood, trans. *Bhagavad-Gita: the Song of God* (New York: Signet Classic: 2002), pp. 110-111.

18 Paramahansa Yogananda, *God Talks with Arjuna: The Bhagavad Git: Royal Science of God-Realization* (Self-Realization Fellowship, 1999), pp. 788-789, emphasis added.

19 Ibid., p. 927.

20 Ibid., p. 928.

21 Ibid., pp. 928-931.

22 Ibid., pp. 71-72, citing the *Purusa Sukta*, the hymn of man, 10:90, emphases added.

23 Ibid., p. 70, emphasis added.

24 Ibid., pp. 148-149.

25 Ibid., p. 153.

26   Joseph Campbell, *The Maskes of God: Volume III: Occidental Mythology* (Penguin, 1991), p. 105.

27   Ibid., p. 108.

28   Ibid.

29   Ibid., pp. 180-181.

30   One need only think of how crucial the spatio-temporal notions of motion and rest became in the theological systems of Origen, or, for that matter, Maximus the Confessor.

❧ Two ❧

THE TOPOLOGICAL METAPHOR OF THE MEDIUM AND ITS REVOLUTIONARY INVERSION:

THE "FIRST EVENT," THE FOUR STAGED DESCENT OF MAN,

AND THE THREE GREAT YAHWISMS

∴

"The one God was the goal of the 'way up,' of that ascending process by which the finite soul, turning from all created things, took its way back to the immutable Perfection in which alone it could find rest."
—Arthur O. Lovejoy[1]

"As an instrument for describing and classifying ancient religions, the opposition of unity and plurality is practically worthless."
—Jan Assmann[2]

WITHIN MOST ANCIENT mythologies and traditions, the Tower of Babel Moment was cast in a wider context in which the creation of man himself was metaphorically described as a descent, a "fall" from a higher position to a lower one. This higher position in turn is a symbol of the primordial Nothingness or Unity Itself - or Himself, as we shall see momentarily - from which all creation emerges in an endless process of differentiations.

One may dismiss, as the contemporary academy or organized monotheistic religions so often do, the multiform mythologies and allegories in which this metaphor has been presented, as a relic of a bygone, less sophisticated and obsolete world view. Or, as we do here, one may assume for the sake of argument and high speculation that it was the legacy of an advanced culture from High Antiquity, whose scientific sophistication was equal to, or exceeded our own. Viewed in that way, the interpretation of ancient myths and texts which contain, *or reject*, that metaphor, changes profoundly.

A. The Metaphor and the First Event,
or Primordial Differentiation
1. The Metaphor and Some of Its Cultural Expressions

But what exactly is this Topological Metaphor of the Medium?

This primordial Nothing and its first differentiation has been expressed under a variety of images and names the world over, but as we shall see here and in a subsequent chapter, it also contains a profoundly sophisticated physics metaphor of the physical medium itself, a metaphor that we call "the Topological Metaphor of the Medium."

To understand it, one need do nothing more than a simple "thought experiment," one that was performed many times by the ancients in their explication of the idea, in this case, by Iamblichus the Neoplatonist, reflecting on the primordial Nothing, which makes its appearance here as "the monad":

The monad is the non-spatial source of number....

Everything has been organized by the monad, because it contains everything potentially: for even if they are not yet actual, nevertheless the monad holds seminally the principles which are within all numbers...[3]

Note that as far as the Metaphor is concerned, the primordial Unity "contains" all in potential, a physics conception that, as we shall discover in a subsequent chapter, is not far from contemporary thought.

But why call this Metaphor a "topological" Metaphor at all? Why reference a higher-dimensional language of mathematics at all? To understand the answer to these questions, it is necessary to reprise what we have written elsewhere about it in connection to the Hindu expression of it.

This implies, however, a "first event," that initial differentiation of Nothing into the first primordial "somethings," an event that may best be understood by elaborating our "thought experiment." As before, imagine a sea of Nothing, infinitely extended in every "direction," though, in a sea of Nothing the idea of direction itself is a thing that really is inapplicable to this Nothingness. One can only speak of this Nothingness analogically, metaphorically.

There is also Nothing "going on" or, to put it in much richer terms, "Nothing is Happening." Since Nothing is Happening, there is no time, since there is no change; All is Sameness; All is Nothing.

One upon a time there was no time at all.

Time is nothing but a measure of the changing positions of objects in space, ad, as any scientist, mystic or madman knows, *in the beginning there were no objects in space*[4].

(Indeed, there was not even "space", since the idea of space, pace Einstein, is inseparable from the objects in it that cause its curvature.)

Despite this initial absence of matter, space and time, something must have happened to get everything started. In other words, *something must have happened before there was anything*.

Since there was noTHING when something first happened, it is safe to say this first happening must have been quite different from the sorts of events we regularly account for in terms of the laws of physics.

Might is make sense to say this first happening could have been in some ways more like a *mental* than a physical event?[5]

Indeed, it is this primordial Unity - the Absolute, the Divine Simplicity, the Grand Architect of the Universe, the primordial androgyny, whatever one wishes to call it - and that First Event, or Primordial Differentiation, that has haunted the minds of mystics, metaphysicians, and, as we shall see in a much later chapter, theoretical physicists, for the millennia that mankind has recorded his thoughts about it.

a. In Hinduism:
(1) The Triune Vishnu

In our previous work *The Grid of the Gods* we unfolded the the various versions of the Metaphor as it occurred in Hindu, Mayan, and Egyptian cosmological philosophy. In the next few sections, we shall quote our previous statements extensively, since it is necessary to perceive the details in order to understand how the Metaphor works. However, we shall also add additional material from the Neoplatonic and Hermetic traditions that may be more familiar to a Western readership. For example, concerning the Hindu version, we began with a citation from the *Padama Purana*:

"In the beginning of creation the Great Vishnu, desirous of creating the whole world, became threefold: Creator, Preserver, Destroyer. In order to create this world, the Supreme Spirit produced from the right side of his body himself as Brahma then, in order to preserve the world, he produced from his left side Vishnu; and in order to destroy the world he produced from the middle of his body the eternal Shiva. Some worship Brahma, others Vishnu, others Shiva; but Vishnu, one yet threefold, creates, preserves, and destroys: therefore let the pious make no difference between the three."[6]

We continued:

Note that neither in the Egyptian nor in the Hindu versions of this "primordial trinitarian homosexual ecstasy" are we dealing with any notion of a theological *revelation.*

We are dealing, rather, with the "topological metaphor" of the physical medium itself, as I noted in the appendix to chapter nine of *The Giza Death Star Destroyed,*[7] and again in *The Philosophers' Stone,*[8]and it is worth recalling what I stated there concerning the emergence of this "trinity" from the information-creating processes of the physical medium as viewed in yet *other* ancient traditions, in this case, the Neoplatonic and Hermetic.

In order to understand what the ancients meant by all the variegated religious and metaphysical imagery they employed to describe this topological metaphor – in order to *decode* it – let us perform a simple "thought experiment." Imagine an absolutely undifferentiated "something." The Neoplatonists referred to this "something" as "simplicity" (απλωτης). Note that, from the *physics* point of view *and from that of Hinduism itself,* we are dealing with a "nothing," since it has no differentiated or distinguishing features whatsoever.

Now imagine one "brackets" this nothing, separating off a "region" of nothing from the rest of the nothing(Vishnu's ejaculation metaphor). At the instant one does so, one ends up with *three* things, each a kind of "differentiated nothing." One ends with:

1) the "bracketed" region of nothing;
2) the *rest* of the nothing; and,
3) the "surface" that the two regions share.

Note something else. From a purely physics point of view, this occurs without *time*, since time is measured only by the relative positions of differentiated things with respect to each other. The "regions of nothing" and their common surface are, so to speak, still eternal, and yet, at the same instant, a kind of "time" has emerged simultaneously with the operation of differentiating itself.

In short, from a non-quantifiable "nothing," information begins to emerge with the process of "bracketing" or "differentiating" itself, including the concept of *number*. On the ancient view, then, numbers do not exist in the abstract. They are, rather, functions of a topological metaphor of the physical medium.[9]

Now let us go further into this topological metaphor by notating our three differentiated nothings mathematically. There is a perfect symbol to represent this "nothing", the empty hyper-set, whose symbol is $\varnothing$, and which contains no "things" or "members." Now let our original "nothing" be symbolized by $\varnothing_E$. A surface of something is represented by the partial derivative symbol ∂, for after all, a "surface" of something, even a nothing, is a "partial derivative" of it. So, we would represent our three resulting entities as follows:

1) the "bracketed" region of nothing, or $\varnothing_{A\text{-}E}$;
2) the *rest* of the nothing, or $\varnothing_{E\text{-}A}$; and,
3) the "surface" that the two regions share, or $\partial\varnothing_{A\text{-}E|E\text{-}A}$.

Note now that the three "nothings" are still nothing, but now they have acquired information, distinguishing each nothing in a *formally explicit* manner from each other nothing. Note something else: *the relationship between them all is analogical in nature, since each bears the signature of having derived from the original undifferentiated nothing; each retains, in other words, in its formal description, the presence of* $\varnothing$. And this will be true *no matter how many times one continues to "bracket" or "differentiate" it.* On this ancient cosmological view, in other words, everything is related to everything else by dint of its derivation via innumerable steps of "differentiation" from that original nothing. It is this fact which forms the basis within ancient civilizations for the practice of sympathetic magic, for given the analogical nature of the physical medium implied by these ancient cosmologies, in purely physics terms, everything is a coupled harmonic oscillator of everything else.[10] Finally, observe how this formal explicitness dovetails quite nicely with the Hindu conception that the created world is, in fact, illusion, a "nothing," but a differentiated nothing.

Now let us take the next step in the decoding of this topological metaphor in ancient texts and cosmologies. It is understood within the kind of mathematical metaphor that we are exploring here, that *functions* can be members of the empty hyper-set without destroying its "emptiness," for the simple reason that *functions* are not "things" or objects, but pure processes. Thus far, we have dealt with regions, and surfaces, now we add *functions*.[11]

In other words, topology is the mathematical language most suited to exhibiting the Metaphor, for its ability to translate the terms of metaphysics - terms which reference a "dimensionless Nothing" - and mythology into a more formal notational symbolism.

(2) the Bhagavad-Gita: The Knower and The Field

There is a further relationship to modern physics contained within the Hindu version of the metaphor, as the following passage from the *Bhagavad-Gita* demonstrates. Here the conversation is between Arjuna, and the Lord Krishna:

Arjuna:

And now, Kirshna, I wish to learn about Prakriti and Brahman, *the Field and the Knower of the Field.*

Sri Krishna:

This body is called the Field, because a man sows seeds of action in it, and reaps their fruits. Wise man say that the Knower of the Field is he who watches what takes place within this body.
    Recognize me as the Knower of the Field in every body. *I regard discrimination between Field and Knower as the highest kind of knowledge.*
    ....
    Now I shall describe That which has to be known, in order that its knower may gain immortality. That Brahman is beginningless, transcendent, eternal. He is said to be equally beyond what it, and what is not.
    ...
    He is within and without; He lives in the live and the lifeless: Subtle beyond mind's grasp; so near us, so utterly distant:

Undivided, He seems to divide into objects and creatures:
Sending creation forth from Himself, He upholds and with-
draws it....

....

You must understand that both Prakriti and Brahman are without beginning.... The sense of individuality in us is said to cause our
experience of pleasure and pain. The jndividual self, which is Brahman mistakenly identified with Prakriti, experiences the gunas which
proceed from Prakriti...

*The supreme Brahman in this body is also known as the Witness. It
makes all our actions possible, and, as it were, sanctions them, experiencing all our experiences.* it is the infinite Being, the supreme Atman. He
who has experienced Brahman directly and known it to be other than
Prakriti and the gunas, will not be reborn, no matter how he has lived
his life.

....

Know this , O Prince,
Of things created
All are come forth
From the seeming union
Of Field and Knower,
Prakriti with Brahman.

....

For, like the ether,
Pervading all things
Too subtle for taint,
This Atman also
Inhabits all bodies
But never is tainted.* [12]

Notably, what emerges in this chapter of the *Bhagavad-Gita* are two crucial
concepts that will inform our subsequent examination of modern physics'
re-casting of the Metaphor: (1) the Observer, called the "Witness," in the
above quotation, and (2) the Field. As will be seen in a later chapter of this
book, the Witness, in this case, mankind himself, is the *sine qua non* of modern theoretical physics' elaboration of the Metaphor of the Medium, and the
Field itself is, much like the ancient Hindu texts, a field of information and
sensation, of *observation*.

b. In Egypt
(1) An Egyptologist Examines the
Akhenaton Monotheist Revolution

German Egyptologist Jan Assmann cites a hymn from the revolutionary period of Pharaoh Akhenaton's version of monotheism in Egypt, where, in spite of the monotheistic tendency Akhenaton wished to pursue, this "one-and-many" dialectic is again in evidence:

> Secret of transformations and sparkling of appearances,
> Marvelous god, rich in forms!
> All gods boast of him
> To make themselves greater with his beauty to the extent of
>   his divinity.

> Re himself is united with his body.
> He is the Great One in Heliopolis.
> He is called Tatenen/Amun, who comes out of the primeval waters
>   to lead the "faces."

> Another of his forms is Ogdoad.
> *Primeval one of the primeval ones*, begetter of Re.
> He completed himself as Atum, being of one body with him.
> He is Universal Lord, who initiated that which exists.[13]

Note the intriguing expression, "primeval one of the primeval ones," a fitting description of the undifferentiated nothing and the resultant analogical "nothings" that follow upon the "first event."

(2) An Esotericist Examines the Traditional Egyptian Cosmology

The eminent esotericist Rene Schwaller de Lubicz added his own interpretation to the Egyptian version of the Metaphor, and once again, it is best to reprise what we said about it previously in *The Grid of the Gods*, for the significance of his understanding of the Metaphor lies in the details:

> The primordial differentiation, which Scwhaller calls the "primary scission," is evident in the Memphite myth, which we may understand as yet another "paleophysical metaphor," i.e., as a profoundly sophisticated physics metaphor disguised in religious terms. There,

the primary scission is, as in the Vedic tradition, expressed in the generation of the gods from the primordial ocean, or Nun:

The revelation of Heliopolis... is the mysterious divine action of the scission of Unity in Nun (the milieu likened to the primordial Ocean), which coagulates into the first earth, incarcerating the invisible fire of Tum.

This is the heavenly fire fallen into earth, which in the mystery of Memphis takes the name Ptah. This metaphysical fire produces its effects in nature by materializing the principles enunciated at Heliopolis, but not as yet manifested.

The appearance of Tum implies the becoming of the three principles and the four essential qualities philosophically called the constituent elements of matter, but their "corporification" takes place only upon the appearance of the first Triad: Ptah, Sekhmet, and Nefertum.[14]

While the emergence of the number four may, at this juncture, seem ad hoc and completely arbitrary, we shall see in a little while that it contains yet another physics metaphor.

For the moment, however, our focus must remain on the emergence of the primordial triad of Ptah, Sekhmet, and Nefertum, for "Immanent in every being is a faculty of numbering that is an *a priori* knowledge of Number. The very fact of distinguishing between the I and the other is an enumeration."[15] In other words, for Schwaller, implicit in the primary scission is its relationship to consciousness and its Unity-in-Diversity. Schwaller explains the primary scission this way:

Thus, at the origin of all creation, there is a Unity that, incomprehensibly, must include within it a chaos of all possibilities, and its first manifestation will be through division. At the origin of all concepts, there is One and Two, being Three principles where one explains the other, incomprehensible in itself.

....

Here is the divine Trinity that is infallibly found at the origin of all things, all arguments and reasoning; the Trinity that supports everything, the foundation on which the world is built, as everything stems from it.

The original Unity contains all possibilities, of *being* and of *nonbeing*. Consequently, it is of androgynous nature.[16]

We have already made reference to this peculiar "primordial androgyny" - the subject of a whole other book - but again, what Schwaller
is pointing out is that in the topological metaphor of the "differentiation of a primordial Nothing," the inevitability of a One-Three always
results: two regions of bracketed nothing sharing a common surface.

Thus we may add the names Ptah, Sekhmet, and Nefertum to
our previous table, indicating a common conceptual inheritance lays
behind Egypt and the Vedic culture:

1) Ptah = $\varnothing_A$;
2) Sekhmet = $\varnothing_B$;
3) Nefertum = $\partial\varnothing_{A,B}$.

...

As already mentioned, why the ancients should so consistently view
this primordial differentiation in androgynous terms is the subject of
another book which we eventually hope to write, but for now it is
worth noting in this regard something else that Schwaller points out:

> "Do you care to translate this as Father, Spirit, and Son or Osiris,
> Isis, and Horus? or Brahma, Siva, and Visnu?
>
> You may, but if you are wise and wish not to be led
> astray, you will say, One, Two, which are Three. This has been
> *represented* by initiates for those who need images, so that they
> may *rally around a tradition*, and be bound by what is called 're-
> ligion.'"[17]

In other words, once one comprehends the fact that the assignation
of various gods' names to the topological metaphor is just that, an
assignation, then one understands that any assertion of a primordial trinity is, in fact, not the consequence of religious revelation or
metaphysics, but a scientifico-philosophical first principle needing
no faith, but rather, a kind of belief in the character of the formally
explicit metaphor, for that metaphor can be described in the highly
abstract symbolisms of topology itself.[18]

Again, the point to be noted is that Schwaller, a mathematician, has understood the highly topological nature of the metaphor, for behind the various
names of the gods lie deeply "higher dimensional" mathematical functions

and concepts, the function of differentiation, and the resulting differentiated "nothings" sharing a common surface, which in itself is also nothing.

c. In Mayan Culture

Again, as we have noted elsewhere, the Mayans too had their own elegant poetic expression of the Metaphor:

Like the Hindu cosmology laid out in stone reliefs at Angkor Wat, The *Popol Vuh* begins in an abyss of mystery, an abyss laid out in eloquent and elegantly simple words and imagery whose power is made even more manifest by their poetic simplicity:

> "This is the account, here it is:
> Now it still ripples, now it still murmurs, ripples, it still sighs, still hums, *and it is empty under the sky.*
> *Here follow the first words, the first eloquence.*
> *There is not yet one person, one animal, bird, fish, crab, tree, rock, hollow, canyon, meadow, forest. Only the sky alone is there;* the face of the earth is not clear. *Only the sea along is pooled under all the sky; there is nothing whatever gathered together. It is at rest;* not a single thing stirs. It is held back, kept at rest under the sky.
> *Whatever there is that might be is simply not there: only murmurs, ripples, in the dark,* in the night. Only the Maker, Modeler alone, Sovereign Plumed Serpent, the Bearers, Begetters are in the water, a glittering light. They are there, they are enclosed in quetzal feathers, in blue-green.
> Thus the name, "Plumed Serpent." They are great knowers, great thinkers in their very being.
> And of course there is the sky, and there is also the Heart of Sky. This is the name of the god, as it is spoken.
> And then came his word, he came here to the Sovereign Plumed Serpent, here in the blackness, in the early dawn.... Thunderbolt Hurricane comes first, the second is Newborn Thunderbolt, and the third is Sudden Thunderbolt.
> So there were three of them..."[19]

By now, this powerful, evocative imagery should recall the image of Vishnu at Angkor Wat, superintending the cosmic tug-of-war of the great *naga* serpent in the Milky Ocean.

Yet, this appears half a world away, in an entirely different culture!

Note too, that the *topological metaphor* of a primordial trinity is preserved. Everything begins as an emptiness "under the sky" and there is not yet any differentiation within it: "there is not yet one person, one animal" and so on. There is only an empty sky, and pooled water at rest beneath it. The only thing existing is Sovereign Plumed Serpent and a mysterious reference to "Bearers" and "Begetters in the water" who are described as "great knowers, great thinkers in their very being," who are later found, just like Vishnu, to be manifestations of Sovereign Plumed Serpent.

The *Popol Vuh* is telling us, in other words, the same thing we saw at Angkor Wat: there is a primordial "nothing", Sovereign Plumed Serpent, and then there is a primordial "trinity," of endless indistinct "sky" and below it a "sea", *and the implied common surface between the two*. Nothing else whatsoever, at this juncture, exists, except a faint "murmuring" and "rippling" in the night, implying somehow that *sound, frequency, vibration* give rise to all the fecund distinctions and variety to follow.

Indeed, at the very beginning, the *Popol Vuh* informs us that "This is the beginning of the Ancient Word, here in this place called Quiché. Here we shall inscribe, we shall *implant* the Ancient Word, *the potential and source for everything done*...in the nation of the Quiché people."[20] Note that the Ancient Word is something to be *implanted*, again recalling the imagery of Vishnu ejaculating into the primordial sea, which was but himself under another manifestation. Note too the very suggestive notion that this Word, this sound or vibration as it were, is "the potential and source for everything done," that is, that all the diversity that arises, arises from this pure and infinite potential.

Consequently, it would appear that the *Popol Vuh*, in its very opening pages, is suggesting the very *same* topological metaphor of the physical medium that we encountered in chapter three, in connection with Vishnu's "trifurcation" and differentiation of himself as a primordial Nothing, and that we also discovered operative in some passages in the *Hermetica*, which were of *Egyptian* provenance, only here the metaphor of that "differentiated Nothing" is even more clearly suggested by the notion of an endless sky and endless sea, in neither of which nothing else exists; there is only the sky, the sea, and the surface touching, differentiating, or bracketing, both; again we have three entities of yet another primordial triad.

...

So now, we may add to what we stated about this topological metaphor of the medium in chapter three, for now we encounter yet more
imagery – sky, sea, and the implied surface between the two – all
saying the same thing: that we are dealing with a differentiated Nothing, whose first differentiation must always be triadic or trinitarian in
nature:

1) the "bracketed" region of Nothing, or $\varnothing_{A-E,}$ Hermes' "Kos-
    mos", the *Padama Purana's* Shiva, and now, the *Popol Vuh's*
    "sky";
2) the *rest* of the Nothing, or $\varnothing_{E-A}$, Hermes' "God," the *Padama
    Purana's* Vishnu, and now, the *Popol Vuh's* "sea"; and,
3) the "surface" Nothing that the two regions share, or $\partial\varnothing_{A-|E-A}$,
    Hermes' "Space," the *Padama Purana's* Brahma, and now, the
    Popol Vuh's implied common surface between "sea" and "sky".

However, the *Popol Vuh* goes on to make an even more interesting
and suggestive set of statements that would seem to associate the creation of mankind itself with this process of emerging differentiation
from some sort of *materia prima* or "primordial nothing."[21]

Before we can summarize the Metaphor, however, it is worth looking at its
more familiar expressions within the Neoplatonic and Hermetic traditions.

d. In Neoplatonic Tradition

For western audiences, the most typical example of this kind of philosophical meditation on the primordial unity in the Metaphor is the system
of Neoplatonism. Let us go all the way back to something that Joseph wrote
in *The Giza Death Star Destroyed* to see how the Metaphor worked in the
principle exponent of Neoplatonism, the philosopher Plotinus.[22]
As in many other such ancient systems, Plotinus' version of the Metaphor expresses itself in the form of an original undifferentiated Nothing,
which he calls "The One"(το εν in the Greek, a term which, interestingly
enough, is neuter in gender). This One, however, gives rise to two further
entities, the Intellect, or Mind (Νους a masculine gendered term in the
Greek), and the World Soul (κοσμικη ψυχη, which is, predictably enough, a
feminine term in the Greek). Thus, again, one has a primordial androgyny
giving rise to a distinction.

But this, for Plotinus, is a secondary consideration, and we shall return to another Neoplatonist, for whom the primordial "alchemosexuality" *is* a primary consideration, a little later.

For our purposes, it is important to note that the Intellect and World Soul constitute a kind of "eternal nature *around* the One (περι το εν). We are dealing, once again, with an original undifferentiated Nothing - the One - giving rise to two further entities. For our purposes, it is important to note that the World Soul derives both from the One and the Intellect; in other words, it functions in much the same way like the common surfaces of the two differentiated "regions of Nothing" we have encountered in other versions of the Metaphor: We have

1) The original Nothing, the One: $\varnothing_A$;
2) The Differentiated Nothing, the Intellect: $\varnothing_B$; and
3) The common surface between the two,
   yet a further differentiation: $\partial\varnothing_{A,B}$.

But there is more to it than just this. As in other versions of the Metaphor, Plotinus understood the One in terms of "simplicity," (απλοτης) a technical term that meant simply that the One possessed absolutely no distinctions, and yet, that simplicity contains potentially all that is.[23] Lacking distinctions between any categories, and yet containing all that is, the One's "First Event" of differentiation means that it is both an act of its will, and yet, a kind of essential act as well (since will and essence are not distinct in the One).

This sets up a tension that one often encounters in various versions of the Metaphor, for one may elect either to understand this "First EventI in an "atheistic" way, as a kind of random act of chance, or one may understand it as the deliberate act of Will and Consciousness. Indeed, in the Metaphor, *both*, strictly speaking, are true, since the primordial Nothing is without distinctions, including those of act, will, essence, necessity, space, or time, randomness, chance, and so on. We will return to the implications for consciousness a little later, for now, there is one final version of this metaphor that must be explored.

e. The High Esoteric Tradition of the Metaphor:
the Hermetica and the Image of Androgyny
(1). God, Space, and Kosmos

It is in the *Hermetica* that we encounter the clearest expressions of the Metaphor, and of man's place within it.[24] Again, we cite what Joseph has written previously about this topic, since the details, again, are important to the argument:

The passage is the *Libellus II:1-6b*, a short dialogue between Hermes and his discipline Asclepius:

"Of what magnitude must be that space in which the Kosmos is moved? And of what nature? Must not that Space be far greater, that it may be able to contain the continuous motion of the Kosmos, and that the thing moved may not be cramped for want of room, and cease to move? – *Ascl.* Great indeed must be that Space, Trismegistus. – *Herm.* And of what nature must it be Aslcepius? Must it not be of opposite nature to Kosmos? And of opposite nature to the body is the incorporeal…. Space is an object of thought, but not in the same sense that God is, for God is an object of thought primarily to Himself, but Space is an object of thought to us, not to itself."[25]

This passage thus evidences the type of "ternary" thinking already encountered in Plotinus, but here much more explicitly so, as it is a kind metaphysical and dialectical version of topological triangulation employed by Bounias and Krasnoholovets in their version in their model. However, there is a notable distinction between Plotinus' ternary structure and that of the *Hermetica*: whereas in Plotinus' the three principle objects in view are the One, the Intellect, and the World Soul, here the principal objects in view are the triad of Theos, Topos, and Kosmos (θεος, τοπος, κοσμος), or God, Space, and Kosmos, respectively.

These three – God, Space, and Kosmos – are in turn distinguished by a dialectic of opposition based on three elemental functions, each of which in turn implies its own functional opposite:

$$f_1: \text{self-knowledge} \Leftrightarrow -f_1: \text{ignorance}$$
$$f_2: \text{rest (στασις)} \Leftrightarrow -f_2: \text{motion (κινησις)}$$
$$f_3: \text{incorporeality} \Leftrightarrow -f_3: \text{corporeality.}$$

So in Hermes' version of the metaphor, the following "triangulation" occurs, with the terms "God, Space, Kosmos" becoming the names for each vertex or region:

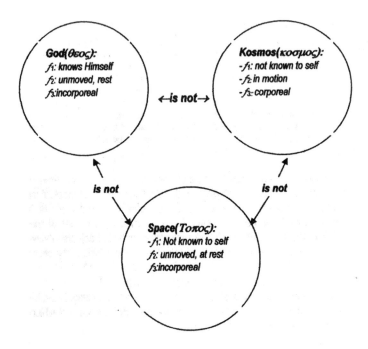

This diagram is significant for a variety of reasons. For one thing, theologically informed readers will find it paralleled in the so-called Carolingian "Trinitarian shield," a pictogram used to describe the doctrine of the Trinity as it emerged in the Neoplatonically-influenced Augustinian Christianity of the mediaeval Latin Church. Again, it must be recalled in this context that the Greek Fathers objected to this formulation of the doctrine in the strongest possible terms, and viewed this dialectical structure as not so much metaphysical, as "sensory," i.e., as more applicable to physical mechanics than to dogmatic theology.

More importantly in this context, however, the diagram illustrates how each vertex – God, Space, Kosmos – may be described as *a set of functions or their opposites:*

God (θεος) $\{f_1, f_2, f_3\}$	Kosmos (κοσμος) $\{-f_1, -f_2, -f_3\}$	Space (τοπος) $\{-f_1, f_2, f_3\}$
f_1: knowledge f_2: unmoved f_3: incorporeal	$-f_1$: ignorance $-f_2$: in motion $-f_3$: corporeal	$-f_1$: ignorance f_2: unmoved f_3: incorporeal

Hermes' version of the metaphor thus lends itself quite neatly to an analysis in terms of Hegelian dialectic, with Space itself forming the synthesis between God, the thesis, and Kosmos, the antithesis, described in terms of the functions f_1, f_2, f_3 or their opposites.

To see how, let us extend the formalism by *dispensing with* Hermes' metaphysical description of the *functions f_1, f_2, f_3* and take the terms God, Kosmos, and Space as the sigils of distinct or discrete topological regions in the neighborhood of each vertex in the diagram on the previous page, and model them as empty hyper-sets. Since it is possible for combinatorial functions to be members of empty sets, then letting $\varnothing_G$, $\varnothing_K$, $\varnothing_S$ stand for God, Kosmos, and Space respectively, one may quickly see the lattice work that results from entirely different sets of functional signatures, exactly as was the case in Plotinus, but via a very different route:

$$\varnothing_G = \{f_1, f_2, f_3\}$$
$$\varnothing_K = \{-f_1, -f_2, -f_3\}$$
$$\varnothing_S = \{-f_1, f_2, f_3\}.$$

Note that space in Hermes' version of the metaphor, since it comprises functional elements derived from the other two regions – "God" and "Kosmos" – could be conceived as the common "surface" between the two. Thus, once again, we have our familiar three entities:

1) the "bracketed" region of nothing, or $\varnothing_{A\text{-}E}$, Hermes' "Kosmos";
2) the *rest* of the nothing, or $\varnothing_{E\text{-}A}$, Hermes' "God"; and,
3) the "surface" that the two regions share, or $\partial\varnothing_{A\text{-}E|E\text{-}A}$, Hermes' "Space."[26]

(2). Androgyny in the Hermetica

Similarly, the *Hermetica* presents the now familiar symbol of "masculine androgyny" as an image of the fusion of differentiations and the potency of all things in God.

For I deem it impossible that he who is the maker of the universe in all its greatness, the Father or Master of all things, can be named by a single name, though it be made up of ever so many others; I hold that he is nameless, or rather, that all names are names of him. For he in his unity is all things; so that we must either call all things by his name, or call him by the names of all things.

He, filled with all the fecundity of both sexes in one, and teeming with his own goodness, unceasingly brings into being all that he has willed to generate... *Asclepius*. You say then, Trismegistus, that God is bisexuual?[27] - *Trismegistus*. Yes, Ascleptius; and not God alone, but all kinds of beings, whether endowed with soul or soulless.... For either sex is filled with procreative force; and in that conjunction of the two sexes, or, to speak more truly, that fusion of them into one, which may be rightly named Eros, or Aphrodite, or both at once, there is a deeper meaning than man can comprehend. It is a truth to be accepted as sure and evident above all other truths, that by God, the Master of all generative power, has been devised and bestowed upon all creatures this sacrament of eternal reproduction, with all the affection, all the joy and gladness, all the yearning and the heavenly love that are inherent in its being.[28]

Note, once again, that the androgyny of God is stated in the oxymoronic form of a *masculine* androgyny, once again raising the question of why - what process of reasoning was employed by - these ancient cultures to employ such an image?

f. Summary of the Metaphor as Examined, and Its Methodological Implications

For the moment, we leave aside this question for a later chapter. For the present moment it will be helpful to gather these versions of the Metaphor - and the reader is reminded that there are many more than those surveyed here - together, before we examine the application of the metaphor to the *social* space in the great inversions of monotheistic religion.

The implications of this sort of analysis are profound and far-reaching, for they suggest that behind certain types of metaphysical texts, particularly those suggesting triadic structures, there is a much deeper topological metaphor that such texts are designed to encode and transmit. It suggests that all such texts are capable of a deep topological analysis, and that they have nothing, really, to do with metaphysics in the conventional philosophical or theological senses at all. They also suggest, as more and more differentiations are added to this process that account for the rise of physical creation, that there is a *physics* reason for the phenomenon of the world grid. They suggest that, as the physical medium is the information-creating and transmuting Philosophers' Stone itself, that the purpose of the world grid and its constructions is one of an "alchemical architecture," of the monumental manipulation and engineering of the medium itself, for after all, on the ancient view, once again, everything derives from that nothing and is a multi-differentiated nothing, directly tied in with everything else.

In these metaphysical and religious texts, in other words, we are looking at a profound topological and physics metaphor. We are looking at declined legacies of a very ancient, and very sophisticated, science.

We are looking also at a metaphor that is common among discrete cultures, from the Mayan to the Hindus to the Hermetic and Neoplatonic traditions. This commonality, viewed from the context of the Tower of Babel moment means that the ancient *philosophia perennis* was a unifying factor, and that it was *also perhaps the threat implied in all versions of the Tower of Babel moment.*

B. The Descent of Man
1. Universe as the Body of God:
Makanthropos, Entanglement and the **Bhagavad-Gita**

In "The Field and Its Knower" from the *Bhagavad-Gita* cited previously, there is a passage that makes it clear that the universe, viewed a certain way within this metaphor, can be understood to be the literal "body" of God, or, if one prefer, the literal "corporification of Nothing":

> *This body is called the Field, because a man sows seeds of action in it,* and reaps their fruits. Wise man say that the Knower of the Field is he who watches what takes place within this body.
> Recognize me as the Knower of the Field in every body. *I regard discrimination between Field and Knower as the highest kind of knowledge.*[29]

This implies that all events or "happenings" within this "body," from the First Event or primary differentiation itself to the very last uttermost happening, are all, like the events that happen to an ordinary body, are connected; all events, in other words, are *entangled.*

The idea that the universe is a kind of body, a kind of living organism or expression of the Supreme Consciousness, also implies, for the ancient Metaphor, that it "was made with humankind in mind,"[30] and that it is a kind of makanthropos or "great man." As we shall discover in a subsequent chapter, there are very modern and sophisticated versions of this view within modern theoretical physics. This in turn implies something very significant: *The ancient metaphor may have itself represent another legacy of a Very High Civilization in High Antiquity, and may thus represent a kernel of scientific truth.*

2. Man as Microcosm

If the universe was "made for man" in the conception of the ancient Metaphor, if it was a "great man" (μακανθροπος) then mankind was in turn a mirror of the cosmos itself; he was a "microcosm" of the universe (μικροκοσμος). This forms the basis of the ancient conception and practice of sympathetic - or as we prefer to call it - analogical magic, for man himself becomes both the medium of operation and the operator or magus himself. Booth sums up the view aptly by stating that the ancients "believed in a quite literal way that nothing inside us is without a correspondence in nature."[31]

3. Mineral, Vegetable, and Animal Man

As the process of differentiations unfold, man descends through four realms, the heavenly, the mineral, the vegetable, and finally enters its current state of existence, the animal. There is, in other words, a "mineral man," a "vegetable man," and an "animal man." In all these prior states, except the last, mankind is perceived as an "androgyny," to such an extent that his reproduction in the vegetable stage, for example, is conceived to be plant-like.[32]

C. THE ESOTERIC TRADITION OF THE PRIMORDIAL UNITY AND ITS SYMBOL IN ANDROGYNOUS MAN

With this in mind, we turn once again to the esoteric tradition and to a reconsideration of the Tower of Babel moment, where, once again, the story is recast in terms of a *fall*, and within a context that to modern sensibilities seems at first absurd and bewildering, but as we shall see in subsequent chapters, actually contains a profoundly modern scientific metaphor.

1. The Primordial "Androgyny" and the Primary Differentiation

Mark Booth makes a cogent observation bearing directly to the nature of our study: "For science the great miracle to be explained is the physical universe. For esoteric philosophy the great miracle is human consciousness."[33] We might take minor issue with Booth's observation, for as we have seen, the essence of the various versions of the Tower of Babel moment is to explain not only mankind's consciousness, but its presumed "original high knowledge" and its current dilemma and weakness. It is thus not merely human consciousness that is in view in the esoteric tradition, but also- via the descent through mineral, vegetable, and animal states - its various states and stages of use.

The famous esotericist Manly P. Hall even noted that the doctrine finds a parallel in rabbinical and esoteric Judaism, and expresses itself in conjunction with an original "masculine-androgyny," or in what we have called "alchemo-sexuality". Hall comments on the esoteric interpretation of Genesis 1:27:

> ... (The) androgynous constitution of the Elohim (god) is disclosed in the next verse, where *he* (referring to God) is said to have created man in *his* own image, *make* and *female*; or, more properly, as the division of the sexes had not yet taken place, *male-female*. This is a deathblow to the time-honored concept that God is a masculine potency as portrayed by Michelangelo on the ceiling of the Sistine Chapel. The Elohim then order these androgynous beings *to be fruitful*. Note that neither the masculine nor the feminine principle as yet existed in a separate state! And lastly, note the word "*re*plenish." The prefix *re* denotes "back to an original or former state or position,", or "repetition or restoration." This definite reference to a humanity existing prior to the "creation of man" described in Genesis must be evident to the most casual reader of Scripture.[34]

The only thing we would take issue with here is with Hall's characterization of Michelangelo's portrayal on the Sistine chapel, for the point of the masculine "alchemosexuality" is not to neglect the feminine, but rather, in our opinion, yet another legacy of the possible scientific basis behind the image, for if it was a legacy of a scientifically created culture, it may have been based on the knowledge that human males carry both sexual chromosomes, and was applied analogically as a metaphor or symbol of the physical medium in which all distinctions were united. This pursuit of analogical thinking with respect to the primordial alchemosexuality even finds expression in the ancient belief that male sperm "was held to contain a particle of the *prima materia* our of which everything was made..."[35] We will explore these physics analogs in the final chapter.

In any case, Hall notes that Judaism has a four-staged descent of mankind, in only the last stage of which mankind was divided into the sexes.[36] Additionally, rabbinical Judaism also conceived of the original masculine-androgyny as a gigantic being whose mass filled the whole world, stretching to all four points of the compass, and that he thus constituted a "future altar."[37] The primordial androgyny of mankind is one of the seedier symbols of the ancient "alchemical vision" of man, but it is important to view it in more than a merely physical sense, but also as a symbol of a state of a more unified consciousness, since it reflects the physical medium directly.

> That is why it is called 'artificer' and 'modeler,' since in its processions and recessions it takes thought for the mathematical natures, from which arise instances of corporality, of propagation of creatures and of the composition of the universe. Hence they call it 'Prometheus,' the artificer of life, because, uniquely, it in no way outruns or departs from its own principle, nor allows anything else to do so, since it shares out its own properties. For however far it is extended, or however many extensions it causes, it still prohibits outrunning and changing the fundamental principle of itself and of those extensions.
>
> So, in short, they consider it to be the seed of all, and both male and female at once - not only because they think that what is odd is male in so far as it is hard to divide and what is even is female in so far as it is easy to separate,[38] and it alone is both even and odd, but also because it is taken to be father and mother, since it contains the principles of both matter and form, of craftsman and what is crafted; that is to say, when it is divided, it gives rise to the dyad.... *And the seed which is, as far as its own nature is concerned, capable of producing both females and males, when scatted not only produces the nature of both without distinction, but also does so during pregnancy up to a certain point; but when it begins to be formed into a foetus and to grow, it then admits distinction and variation one way or the other, as it passes from potentiality to actuality.*[39]

This is, we must admit, one of the most disconcerting passages and applications of the androgyny symbol we have ever come across, for consider the implications. While it is true that the idea of foetal androgyny and subsequent distinction can arise from the standard Greek doctrines of form(the androgyny itself) and differentiation (the distinction of male and female), the passage clearly implies some sort of knowledge that a foetus during the first weeks of pregnancy has characteristics of both sexes. As we shall see in a later chapter, this could also be construed, not as a rationalization from the doctrine of form and matter, but as a legacy of actual scientific knowledge handed down from a more sophisticated scientific culture, for it is indeed the case that during the first weeks of pregnancy, a human foetus has the rudimentary sex organs of both sexes.

This, of course, contains its own thorny implications, for if that be the case, then the selection of androgyny as a symbol of the primordial unity and descent of man *may* have been carefully chosen on the basis of - and to convey - a more scientific knowledge based in the "embryonic androgyny."

Thus the high esoteric tradition presents us with something of an inconsistency, a logical paradox, for at the top of the hierarchy, lies the Undifferentiated Nothing, a kind of "primordial androgyny," a symbol for the undifferentiated Unity in which all concepts find a synthesis, known under various names - the Absolute, the Grand Architect, the One, the All, the Unknowable - beneath Him, as the differentiations of the metaphor "descend," come the Mineral, Vegetable, and finally, the Animal worlds. The paradox lies in the fact that the animal world, the realm of life as we know it, occupied both the lowest level, and yet, is the level that all others were done in preparation for. Oddly, this basic cosmology is once again in the same broad progression that is maintained by modern physics cosmology and biological evolution, suggesting once again that these images were possibly inspired by a scientifically sophisticated culture and civilization.

There are thus, within the esoteric version of the primary "event" and its subsequent differentiations, "four Adams" or four humanities, the Mineral, the Vegetable, and finally, the Animal, the "Adam" or humanity that we are now.

2. Its Implications

If this "topological metaphor" of the physical medium be true, then several implications immediately follow:

1) the metaphor may be interpreted "atheistically" or "theistically," for in either case, the physical medium is a Nothing that differentiates itself; it is both One and Many, both an All-Consciousness and several individual consciousnesses;

2) it thus needs no spokesmen or institutions, nor can there be distinctive "special revelations", for every human being is quite literally a direct manifestation and expression of it. However, once one under-

stands this, then one can understand why such a philosophical underpinning could produce various social orders, from the Sumerian notion that kingship literally descended from heaven, along with the gods and man themselves, to the later Greek democracies or even Plato's *Republic*, in which every person contributed to the social order, since each was a manifestation of the differentiations of that primordial Nothing; and thus,

3) there can be no system of a uniquely religious revelation;

4) Man's Fall is a result of a movement *from* lesser toward greater differentiation, and thus, Man's fall can be repaired by Man himself, via techniques and technologies designed to "reascend" along the path of descent.

It is *this last* possibility that informs the various components of that technological ascent, and the social, political, and spiritual implications that they entail, and it is this possibility and those entailments that are the subject of the rest of this book.

3. The Inverted Implications:
The Three Great Yahwisms and the
Struggle Against the Prisca Theologia

When viewed against this backdrop of the Topological metaphor, the great monotheistic religions appear as nothing less than a revolutionary world view deliberately designed to overturn the old unifying order in yet another, very hidden, Tower of Babel moment; they appear, in a certain sense, as usurpations, for they change completely the ordering of the cosmos and the community of man within it. For our purposes, we shall not focus on Jainism, Sikhism, or Buddhism, even though they may with some justification be qualified as monotheisms. Rather, our focus is upon the great Monotheisms that are all somehow tied to the character, text, and institutions of the biblical god, Yahweh. These we shall call simply "The Three Great Yahwisms."[40]

a. The Inversion of the Topological Metaphor
to a Technique of Social Engineering
and Construction Via Conflict

The first revolution - and usurpation - against the old order that Yahwism introduced is the alchemical transformation and overturning of the primary scission or first event as a *cosmological principle* and its transmutation into a principle of social engineering, for if the ancient metaphor can be described by a topology of the differentiation of a primordial Nothing into two distinguished regions sharing a common surface, the same could easily be applied, by a kind of inversion, to the mass of humanity. In this application or construction of the Metaphor, the mass of humanity becomes the original undifferentiated medium, and the differentiation is the result of the Yahwist revelation itself, which distinguishes between the original mass and "the chosen people."[41] To put it succinctly, the acceptance of a "special revelation" given to a certain group by a certain individual or sanctioned spokesman or institutions, carries with it certain inevitable social consequences, the first of which is the division of human "social space" into two distinct regions, one of "truth" and the other of "error." It is the first of many such alchemical transformations in the social order.

To put it more succinctly still, the Three Great Yahwisms - Judaism, Christianity, and Islam - are all alchemical techniques of the social engineering of mankind into permanent division.

There is, however, more than just a rupture of the social space which these revelations engender. There is also a rupture in the *temporal* order, in the order of historical and cultural memory, and of "future expectation" when the relvelation and chosen people are vindicated. The Yahwist revelation

> ...was therefore a radically new distinction which considerably changed the world in which it was drawn. The space which was "severed or cloven" by this distinction was not simply the space of religion in general, but that of a very specific kind of religion. We may call this new type of religion "counter-religion" because *it rejects and repudiates everything that went before* and what is outside itself as "paganism." It no longer functioned as a means of intercultural translation; on the contrary, it functioned as a means of intercultural estrangement.[42]

Exactly as was seen in the original context of the Metaphor, where the state of the system after the First Event results in a kind of "break in the space and with the past", the same now happens in the social order.

Consequently, Yahwism inevitably produces the conditions of conflict, and hence, empowers the elite introducing and maintaining it, and this "cloven space" and revolutionary break with the religious past of the rest of mankind becomes a kind of permanent "Tower of Babel Moment," a further fragmentation of man's unity by ideological means, for its primary symbol - the Exodus of Israel from Egypt - becomes a powerful symbol of this perpetual conflict.

Israel embodies truth, Egypt symbolizes darkness and error. Egypt loses its historical reality and is turned into an inverted image of Israel. Israel is the negation of Egypt, and Egypt stands for all that Israel has overcome.[43]

Thus, Yahwism is not a normal progression, or "further unfolding of meanings latent and inherent in the Topological Metaphor;" it is rather a vast social perversion of it, a twisting of it into an alchemical technique of social construction:

Monotheistic religions structure the relationship between the old and the new in terms not of evolution but of revolution, and reject all older and other religions as "paganism" or "idolatry." Monotheism always appears as a counter-religion. There is no natural or evolutionary way leading from the error of idolatry to the truth of monotheism. This truth can come only from outside, by way of revelation. The narrative of the Exodus emphasizes the temporal meaning of the religious antagonism between monotheism and idolatry. "Egypt" stands not only for "idolatry" but also for a past that is rejected.[44]

And lest the alchemical operations are not perceived, we highlight them in the next passage:

The Exodus is a story of emigration and *conversion*, of *transformation* and renovation, of stagnation and progress, and of past and future. Egypt represents the old, while Israel represents the new. The geographical border between the two countries assumes a temporal meaning and comes to symbolize two epochs in the history of humankind. The same figure reproduces itself on another level with the opposition between the "Old" and the "New" Testaments. Conversion presupposes and constructs an opposition between "old" and "new" in religion.[45]

In other words, Yahwism as a social construct was deliberately designed to further enhance the fragmentation of mankind recorded in its own version of the Tower of Babel Moment, for now, in addition to the linguistic fragmentation, the natural knowledge that posed such a threat to the gods is buried beneath a further layer of *ideological fragmentation and cognitive diversion*, this time by the special "knowledge" that comes via the special revelation of monotheism itself. The distinction of Unity and Plurality that was upheld in a both-and dialectical construction in the Topological Metaphor has now undergone a transformation by a kind of alchemical inversion into an *opposition* of Unity and Plurality that is manifest by the social and religious antagonism of Egypt, representing plurality, "polytheism" and "natural religion", and Israel, representing unity, "monotheism" and a "knowledge" only given by revelation.[46]

The result, of course, has been endless conflict, not only between the Three Great Yahwisms and the empowered elites that maintain its authoritative pronouncements, but also to endless conflicts and divisions *within* each of them, between Protestant and Catholic, or Catholic and Orthodox, or Orthodox and Reformed Judaisms, or Sufi and Shia Islam. The "space" and cultural "time" this engendered by this transformation of the Metaphor into a Technique of Social Engineering has thus resulted in the creation of a schizophrenic cultural space within the Western and Islamic worlds.[47]

There is a further implication as well, for once one admits the possibility that the Topological Metaphor - a *physics* construct - may be applied as a basic alchemical *technique of the social transformation of man*, then one admits of the possibility that all techniques and technologies have implications for , and may be applied to, the engineering of humanity and its culture and society itself. In other words, one admits the possibility that each of the levels of the descent of Man in the esoteric cosmology may become (and indeed are) appropriate goals for the alchemical "reascent" and apocalyptic re-engineering of man.

b. Monotheism and the Resulting Social Dualism: The Convert-Enemy Paradigm of Social Interaction

This schizophrenic cultural "space-time" is to a certain extent also the result of the unique characteristics of Yahwism. Egyptologist Jan Assmann summarizes these features in the following fashion:

> It seems evident that all founded or, to use the eighteenth-century term, "positive" religions are counter-religions. This is so because all of them had to confront and to reject a tradition. None of them was founded within a religious void. Therefore, they may be termed "secondary religions" because they always presuppose the preceding and/or parallel existence of "primary religions." We have no evidence of evolutionary steps leading from primary to secondary religions. Wherever secondary religions occur, they always seem to have been established by foundational acts such as revolution and revelation. Such positive acts often have their negative complements in rejection and persecution. "Positive" religions imply negated traditions.

....

> Secondary or counter-religions are determined and defined by the distinction they draw between themselves and primary religions.

...

> ... The reason for this difficulty is that there seems to exist a necessary link between counter-religions and canonization. All counter-religions base themselves on large bodies of canonical texts. First of all, counter-religions, or secondary religions, appear in textual space, that is, in the form of textual articulation and scriptural tradition, as a specific kind of collective memory based on richly structured textual architectures, inherited and kept alive by means of elaborate techniques and institutions of interpretation.... The distinction between primary and secondary religions appears always as the distinction between nature and Scripture.[48]

As was seen previously, the Topological Metaphor carried the implication that no special individual, text, elite group, or institution uniquely embodied or could lay claim to a special relationship with the physical medium with final and unquestionable authority, for in its "both-andness", in its Unity-in-Diverisity construction, potentially all diversity, all individuals, all cultures, were expressions of it. Its book was nature, and its ritual was, to varying degrees, the topological Metaphor and the "analogical magic" that this engendered, be it the actual practice of temple ceremonials to the simple practice of contemplation or meditation.

In contrast, the Three Great Yahwisms perform a vast series of alchemical inversions to this order:

1) They constitute themselves as a rupture in the "social space" of mankind, by defining themselves in opposition to the rest of it by dint of the possession of a unique truth *for which the claim to universality is made; thus, viewed in terms of the topological metaphor, a derivative space is elevated to be the universal one;*

2) They thus constitute themselves as a break with the past as well as a rupture in the social space, yet, by so defining themselves in opposition to a larger space and previous cultural tradition, they are always in an interior state of cultural schizophrenia, a kind of dysfunctionality that lends itself to exploitation and manipulation;

3) This internal schizophrenia is manifest in the exterior social dualism between the community of "truth" versus those outside it;

4) This special "truth," however, is not due to some unique or new insight into the unfolding of the implications of the Metaphor, but is rather by dint of the revolutionary revelation itself, and expresses itself in the dichotomy of Nature versus Scripture, or Nature versus Scripture's interpretive magisterium;

These implications carry further consequences, not the least of which is the "Convert-Enemy" paradigm of social interaction, for in a certain sense, *the Yahwist monotheism is really a dualist political revolution* in the context of a surrounding culture of Monism, understanding that monism to embrace the Unity-in-Diversity paradigm.

The social dualism inherent in Yahwism carries the important consequences of the "convert-enemy" paradigm, that is, by elevating one social space to the status of "truth," it performs a threefold alchemical transformation of that portion of humanity accepting it:

1) It elevates that special claim to "truth" to the status of a universal claim, and thus engenders the programmatic response that everyone who does not accede to that truth is viewed either as a potential convert, or, failing conversion, an enemy, a heretic, an infidel, with the result that total conversion or conflict with that "otherness" becomes a permanent feature of the construct;

2) It introduces into the cultural construct of that particular social space the idea of a *binary* logic based on the distinction of the "One True God" versus "the Many false gods;"[49] in short, it replaces the *triadic* structure of the logic of the Topological Metaphor with a *binary* one whose implications, again, are perpetual conflict;[50]

3) It introduces the construction of moral opposition, i.e., it construes distinctions as morally opposed constructs.

Consequently, in terms of the Metaphor being developed here, the name "Yahweh" appears as a symbol of the function of binary dialectical opposition, particularly at the cultural and sociological level. It is important to understand the dynamic of this last point correctly. While there was certainly conflict - bloody and huge conflict - between societies that adhered to the *prisca theologia* or the ancient theology, this conflict was not an inevitable and logical outgrowth of that religious matrix itself. Indeed, to a certain extent it might be said that conflict was logically antithetical to it.

With the Three Great Yahwisms, however, conflict is inherent to the system itself:

For these religions, and for these religions alone, the truth to be pro-claimed comes with an enemy to be fought. Only they know of her-etics and pagans, false doctrine, sects, superstition, idolatry, magic, ignorance, unbelief, heresy, and whatever other terms have been coined to designate what they denounce, persecute and proscribe as manifestations of untruth.[51]

The conflict, in other words, engineers a social dualism out of the very nature of the revolutionary claims of these religions, a conflict that, as we shall dis-cover in a moment, implies its own final alchemical apocalypse, as the meta-physical reascent to Unity is transformed once again into a *social* goal and agenda: the complete conversion of all of mankind to its particular "truth."

c. Nihilism as the Distinguishing Characteristic of Yahwism

This "convert-enemy" paradigm carries with it not only the dynamic for perpetual conflict and fragmentation, but also another, much more dangerous, impulse, that of Nihilism: the annihilation of the Other, either by annihilating the "truth" of the Other through conversion and rupture with its past, or by the actual conquest and elimination of it. As was seen previously, the both-and dialectic of the Topological Metaphor, and its ability to cast the First Event as a primordial Diversity-in-Unity, was an element - and indeed the *principal* element - of the metaphor in almost all cultures expressing it. Thus, "As an instrument for describing and classifying ancient religions, the opposition of unity and plurality is practically worthless. God's oneness is not the salient cri-terion here but the negation of 'other' gods."[52] Indeed, to a certain extent, the classification of religions as "monotheistic" or "polytheistic" is a modern con-trivance, and very much the result of the Three Great Yahwisms themselves:[53]

...(The) original meaning of this idea is not that there is one god and no other, but that alongside the One True God, there are only false gods, whom it is strictly forbidden to worship. These are two different things. Asserting that there is only one god may be quite compatible with accepting, and even worshipping, other gods, so long as the relationship between god and gods is understood to be one of subordination, not exclusion. Exclusion is the decisive point, not oneness.[54]

As we saw, subordination was inherent in the Topological Metaphor at the outset, and by the same token, *so also was inclusion*. In the Three Great

Yahwisms, however, one of the derived regions has elevated itself to claim to be the original Undifferentiated state, or to speak exclusively for it. Viewed in this context again, the Metaphor would unequivocally condemn these religions as false since they overturn the nature of the Metaphor itself, and moreover, *would appear to be designed to do so.* In this sense then, one may view Islam as the final logical evolutionary step in this development, and Mohammed as the final and logical prophetic expression of it, for it is there that absolute Oneness and exclusion, with all its nihilistic consequences, have been most clearly refined and expressed.

d. The Binary Logic of Yahwism Versus the Triadic Logic of the Metaphor and the Alchemical Eschatological Necessity

The introduction of the category of "true-false" binary logic into religion via the Three Great Yahwisms has a parallel within science and the rise of Greek Rationalism:

> Science's intolerance or potential for negation is expressed in two directions: in its capacity to distinguish between nonscientific and scientific knowledge, on the one hand, and between false and correct scientific knowledge on the other. Myths are forms of nonscientific knowledge, but they are not for that reason erroneous. Scientific errors are instances of disproved scientific knowledge, but they are not for that reason mythic. We find something similar when we look at counterreligions. Primary religions are "pagan," but they are not for that reason heretical; heresies are heterodox opinions and practices, but they are not for that reason primary religions, nor are they pagan.
>
> The analogy between religion and science... could be spun out much further. But more is at stake here than a mere analogy. The new concept of knowledge has as its corollary that it defines itself against an equally new counterconcept, that of "faith." Faith in this new sense means holding something to be true that, even though I cannot establish its veracity on scientific grounds, nonetheless raises a claim to truth of the highest authority. Knowledge is not identical to faith, since it concerns a truth that is merely relative and refutable, yet nonetheless ascertainable and critically verifiable; faith is not identical to knowldege, since it concerns a truth that is critically nonverifiable, yet nonetheless absolute, irrefutable, and revealed. *Prior to this distinction, there existed neither the concept of knowledge that*

is constitutive for science nor the concept of faith that if constitutive for revealed religion.... The ancient Egyptians, like all other adherents of primary religions, knew about the gods rather than believing in them, and this knowledge was not defined in terms of "true and false," but allowed statements that, to our eyes, seem to contradict each other to stand side by side.[55]

Put differently, the Topological Metaphor did not require faith in the religious sense, but rather, a belief in a formal proposition or presupposition, in a mathematical "given," as the initial postulate of a system from which certain deductions or inferences were made.

Yahwism, on the other hand, required a belief in that proposition and a *faith in its inversion, namely, that one person revealed himself, or was given a formal revelation, to speak in behalf of that primordial Unity directly, exclusively, and without any possibility of appeal.* Again, we see that with the rise of the Three Great Yahwisms, there is a revolutionary rupture with the cultural time that preceded it, with the culture of the Topological Metaphor. This was no longer a cultural space and time of Unity-and-Diversity, this was a cultural space and time of the *opposition* of Unity *to* Diversity.

The problem this poses for these conventional religions, as science advances and seems to be playing out the themes of the original mythologies that expressed the Metaphor, is thus an acute one, for sooner or later, adjustments to them will have to be made in the form of further "final revelations" to encompass the unfolding possibilities, or they will have to be rendered obsolescent by a series of events engineered from their premises so horrific in nature that humanity will not return to them as sources of spiritual and religious truth. Science, too, however, appears to be locked in the same cultural schizophrenia that the Yahwisms introduced into modern culture, pursuing technical and technological fulfillments of the creation of things once only conceived in the mythologies of the Metaphor.

This apocalyptic implication may not, however, be fully understood without understanding the "final universal triumph" that the great monotheisms inevitably imply, for each has its own version of its own final end-time victory over all falsehood: Judaism expects its Messiah, certain strains of Christianity expect the return of Christ to earth and the final earthly triumph of "the kingdom of God," and Islam similarly expects a final universal triumph under its "messiah," the Imam Mahdi. But it is important to understand just what this means: it is nothing less than the assertion of a "final nihilism," the final annihilation by conversion or conquest of all that does and all who do not conform to this "revealed" and "absolute" truth. It is worth noting, too, that as the founding events of these monotheisms were in violence, so too are most of their versions of the "final nihilism."[56]

But as the Metaphor implies its own agenda of reascent, so too these final universal triumphs, like the religions that spawned them, could equally be the staged or engineered events for an apocalyptic and alchemical transformation of man. Those, however, are the subject for another book, for now, our task of surveying the Metaphor is concluded, and the task of outlining its applications via alchemical techniques and technologies of human transformation, now begins. However, first, it will be helpful to review the conclusions and implications of what we have discovered thus far.

ENDNOTES

1   Arthur O. Lovejoy, *The Great Chain of Being*, (Harvard University Press, 1964), p. 83.

2   Jan Assmann, *The Price of Monotheism* (Stanford University Press, 2010), p. 31.

3   Iamblichus the Neoplatonist, *The Theology of Arithmetic*, p. 35.

4   Mark Booth, *The Secret History of the World as Laid Down by the Secret Societies* (Woodstock: The Overlook Press, 2008), p. 20, all emphases in the original.

5   Ibid., all emphases in the original.

6   W.J. Wilkins, *Hindu Mythology* (New Delhi: Heritage Publishers, 1991), p. 116, citing the *Padama Purana*.

7   Joseph P. Farrell, *The Giza Death Star Destroyed* (Kempton, Illinois: Adventures Unlimited Press, 2005), pp. 222-245.

8   Joseph P. Farrell, *The Philosophers' Stone: Alchemy and the Secret Research for Exotic Matter* (Feral House, 2009), pp. 43-48.

9   The similarity of this concept to Schwaller De Lubicz's understanding of numbers in ancient Egypt as *functions of geometry* is readily apparent. Schwaller, a mathematician, knew that he could have expressed this conception more deeply, in the form of numbers not as functions of geometry, but of an even higher-order, as functions of the topology of the physical medium itself. It is my opinion that he did not do so, not because he was unaware of it, but rather, because he was trying to popularize and render Egyptian cosmological thought understandable to lay audiences.

10   Of course, everything is not necessary an *efficient* oscillator of any other given thing, but that is a more complex aspect of the ancient cosmologies and their topological metaphor than can be explored in this chapter. That is the purpose of the rest of this book.

11   Joseph P. Farrell and Scott D. de Hart, *The Grid of the Gods* (Adventures Unlimited Press, 2011), pp. 71-73.

12   *Bhagavad-Gita,* Ch 15, "The Field and Its Knower," Trans. Swami Prabhavananda and Christopher Isherwoodi, pp. 100-105, emphasis added.

13   Hymn 200, Jan Zandee, *Der Amunshymnus des Pap. Leiden I 344vso,* 3 Vols., (Leiden: Instituut voor het Nabije Oosten, 1992), no. 138, cited in Jan Assmann, *Moses the Egyptian*, p. 196, emphasis added.

14   Schwaller de Lubicz, *The Egyptian Miracle*, p. 41.

15   Ibid., p. 43.

16   Ibid., p. 75, emphasis in the original.

17 Ibid.,, pp. 76-77, emphasis in the original.

18 Joseph P. Farrell and Scott D. de Hart, *The Grid of the Gods*, pp. 282-285.

19 *Popol Vuh*, trans. Dennis Tedlock, pp. 64-65, emphasis added.

20 Schwaller de Lubicz, op. cit, p. 63, emphasis added.

21 Joseph P. Farrell and Scott D. de Hart, *The Grid of the Gods*, pp. 181-183.

22 Q.v. Joseph P. Farrell, *The Giza Death Star Destroyed* (Adventures Unlimited Press, 2005), pp. 222-231. The analysis presented in *The Giza Death Star Destroyed* is slightly different than that presented here, and it is to be noted that we regard this version as a more developed reflection on the Metaphor in Plotinus.

23 See the discussion in the "Introduction," by Joseph P. Farrell, to Saint Photios, *The Mystagogy of the Holy Spirit* (Holy Cross Orthodox Press, 1987), pp. 20-21.

24 We are not here concerned with the issue of the authorship or provenance of the *Hermetica,* though we readily grant that the writings attributed to "Hermes Trismegistus" we written ca. 200AD or later, we also believe that at least some of the elements contained within them stem from ancient Egyptian religious cosmology, as well as the obvious Platonic and Neoplatonic influences.

25 *Libellus: 1-6b, Hermetica,* trans. Walter Scott, Vol. 1, pp. 135, 137.

26 qv. Joseph P. Farrell and Scott D. de Hart, *The Grid of the Gods,* pp. 73-77. Also see Joseph P. Farrell, *The Giza Death Star Destroyed,* pp. , Joseph P. Farrell, *The Philosophers' Stone: Alchemy and the Secret Research for Exotic Matter* (Feral House), for other context of these observations.

27 The Latin here is "*Asclepius.* Utriusque sexus ergo deum dicis, o Trismegiste? - *Trismegistus.* Non deum solum, Aslclepi, sed omnia animalia et inanimalia."

28 *Asclepius III: 20b-21, Hermetica,* Volume I, trans. Walter Scott (Kessinger Publishing Company, No Date), pp. 333, 335.

29 "The Field and its Knower," *Bhagavad-Gita,* Swami Prabhavananda and Christopher Isherwood, trans., p. 100.

30 Booth, op. cit., p. 23.

31 Ibid,., p. 39.

32 Ibid., p. 60.

33 Ibid., p. 29.

34 Manly P. Hall, *The Secret Teachings of all Ages,* Readers' Edition, pp. 399-400, all emphases Hall's. These ideas find further expression within rabbinical tradition. For example, the Babylonian Talmud, Negilla 9a glosses Genesis 1:27 as follows: "A male with corresponding female parts created He him." (See Wayne A. Meeks, "The Image of the Androgyne: Some Uses of a Symbol in Earliest Christianity," *History of Religions*, Vol. 13, No 3 (Feb. 1974) pp. 165-208, p. 185, n. 88.)

35 Booth, op. cit., p. 99.

36 Hall, op. cit., p. 401.

37 Ibid.

38 For the maleness of odd numbers and femaleness of even numbers, in relation to ancient musical theory, see our *Grid of the Gods* (Adventures Unlimited Press, 2011), pp. 237-240.

39 Iamblichus the Neoplatonist, *The Theology of Arithmetic,* trans. Robin Waterfield, p. 38, emphasis added.

40 We are following the suggestion of Egyptologist Jan Assmann, *The Price of Monotheism*: "One should therefore speak more properly of a monoyahwehism, as is clearly expressed in the formula *JHWH echad* in the Shema prayer. Yahweh is unique, the one god to whom Israel binds itself." (p. 39.)

41 Jan Assmann in his *Moses the Egyptian* begins his book with a citation of George Specer Brown's *The Laws of Form,* which itself contains a mathematical and topological statement of the First Event as the necessary and logical first step in the chain of virtual events creating mathematical forms. From there, Assmann moves immediately to an application of that law to the construct of Yahwistic monotheism:

"Draw a distinction."

"Call it the first distinction."

"Call the space in which it is drawn the space severed or cloven by the distinction."

"It seems as if George Spencer Brown's "first Law of Construction" does not apply solely to the space of logical and mathhematical construction. It also applies surprisingly well to the space of cultural cohnstructions and distinctions and to the spaces that are severed or cloven by such distinctions."(*Moses the Egyptian*[Harvard Universikty Press, 1997], p. 1, citing George Spencer Brown, *The Laws of Form*(New York: The Julan Press, 1972), p. 3.

42 Assmann, *Moses the Egyptian*, p. 3, emphasis added.

43 Ibid., p. 7.

44 Ibid.

45 Ibid, emphasis added.

46 Assmann states the point this way: "The counter-religious antagonism was always constructed in terms of unity and plurality. Moses and the One against Egypt and the Many. The discourse on Moses the Egyptian aimed at dismantling this barrier. It traced the idea of unity back to Egypt." (*Moses the Egyptian,* p. 168).

47 See Assmann's comments on p, 2ff.

48 Ibid., pp. 169-170.

49 Jan Assmann, *The Price of Monotheism*, p. 2. Assmann states this point as follows: "What seems crucial to me is not the distinction between the One God and many gods but the distinction between truth and falsehood in religion, between the true god and false gods, true doctrine and false doctrie, knowledge and ignorance, belief and unbelief." (p. 2)

50 This carries with it profound interpretive implications, for it would mean that the Christian Trinity, and the many systems of esoteric Judaism and Islam, are attempts within those religions to restore that which was lost in the monotheistic revolution itself. Another way of saying this is that those attempts stem not from their special revelation, but from the older esoteric or rejected tradition.

51 Assmann, *The Price of Monotheism*, p. 4. See also pp. 11, 18.

52 Ibid., p. 31.

53 Ibid. It should be noted that there have been recent scholarly studies that indicate that the Hebrew monotheism actually embraces not only the idea of one supreme god but also a "council" of gods.

54 Ibid., p. 34.

55 Ibid., pp. 14-15, emphasis added.

56 Ibid., p. 22.

❧ Three ❧

THE ALCHEMICAL AGENDA OF THE APOCALYPSE:
CONCLUSIONS TO PART ONE

❖

"...(Alchemy) had always been associated with the idea of time and timing, and that, as Fulcanelli informed us, chiliasm lay at the center of the idea of transforming time itself."
—Jay Weidner and Vincent Bridges[1]

THE TOWER OF BABEL MOMENT and the "Topological Metaphor" of the physical medium - with all its rich variety of imagery presented in religious and philosophical traditions from around the world - constitute the road-map for the alchemical agenda openly hidden in modern science and advocates of "transhumanism."

We have seen, for example, that the Tower of Babel Moment made certain assumptions about mankind:

1) Its initial unity, which expresses itself in three primary ways:
   a) As a sexual, or androgynous unity;
   b) As a linguistic unity; and finally, and perhaps most importantly,
   c) As a cultural-philosophical unity. In this instance, the "Topological Metaphor" also reveals the fact that this ancient philosophy was exactly what the Mediaeval and Renaissance Hermeticists claimed it was, the *prisca theologia*, the "ancient theology."
2) This unity in all its facets constituted some sort of threat to the gods or God, and had to be broken. Notably, when one looks at *all* the ancient records, the unity was indeed broken *at each of the three levels noted above.*

The Tower of Babel Moment also presents its own unique metaphor of the Fall of Man, expressed as a *topological descent* of ever-increasing differentiations, from the primordial androgyny of man, an image and symbol of the "androgyny" of the physical medium, or of God, itself, through the mineral, vegetable, and finally, animal man.

This esoteric component is crucial to an understanding of yet another identifying marker of the alchemical agenda for the transformation of man, for it gives its modern exponents, the transhumanists, access to the ancient doctrine of the *philosophia perennis,* namely, that man is the microcosm of the universe. Thus, *to dominate the universe one must dominate man, and to dominate man one must in turn dominate and control the input to his senses: sight, hearing, touch, smell, and taste, and finally, his sexuality and consciousness themselves.* As we shall discover in the final chapters of this book, this very ancient conception of Man as the Microcosm finds astonishing confirmation in some aspects of modern theoretical physics.

If, therefore, one is to detect an alchemical or hermetic agenda within modern technology or science, one must look for activity in the following areas:

1) domination of *sight* through all the media of that sensory input: art, architecture, and most importantly, any media of *information* assimilated through sight, i.e., literature,[1] and more lately, film and television;

2) domination of *sound* through music and speech;

3) domination – and this is a crucial point – of man's taste input, his *food*, an activity particularly distinguishing of the Anglo-American elite in general, and as we shall see, of the Rockefeller family and its allies in particular;

Additionally, one must look for this activity being exhibited in all the levels of the re-ascent of man according to the basic outlines of the ancient doctrines, from animal, to vegetable, mineral, and finally, androgyny. *To put it differently, one must climb back up, one must re-ascend, the path that led to the current differentiation, from animal, to vegetable, to mineral, and finally, to androgynous, man.*

To summarize this attitude, we may call it *"the full spectrum dominance of man the microcosm."* This is true whether we are speaking of the national elites of particular countries, or the transnational elites of multinational corporations and banks.

As we shall now see in the remainder of this book, we believe that transhumanism is but a the old hermetic alchemy in the new clothes of scientific techniques, but the goals remain the same, for the techniques and technologies are essentially those of social engineering - of the alchemical transformation of man - but the claim of the transhumanist alchemist is that their scientific peregrinations will enhance humanity by vast expansions of consciousness and knowledge, when as we shall argue, it may actually inhibit the advancement of knowledge.

What is new, in other words, is merely the scientific and engineering techniques, but the goals remain very old.

And that may in fact be the design...

ENDNOTES

1   Jay Weidner and Vincent Bridges, *The Mysteries of the Great Cross of Hendaye: Alchemy and the End of Time* (Rochester, Vermont: Destiny Books, 2003), p. 38.

II.

THE NEW FRANKENSTEINS:
THE TRANSGENIC TRANSFORMATION OF MAN
AND THE ALCHEMICAL ASCENT FROM ANIMAL,
TO VEGETABLE, TO MINERAL MAN

---

*"The present age ... prefers the sign to the thing signified, the copy
to the original, fancy to reality, the appearance to the essence ... for
in these days illusion only is sacred, truth profane."*
—Ludwig Feuerbach

"Man is what he eats"
—Ludwig Feuerbach

❧ Four ❧

OLD HOMUNCULI AND NEW FRANKENSTEINS:
GENETICS, CHIMERAS AND THE CREATION
OF "ALCHEMANIMAL" MAN

∴

"...(We) live in the era of 'Frankenfoods,' cloning, in vitro fertilization, synthetic polymers, Artificial Intelligence, and computer generated 'Artificial Life.'"
—William R. Newman[1]

HERMETICISM: there can be no doubt that it exercised an extraordinary degree of influence on the emergence of modern science, an influence many choose to forget or to bury. The extent to which this is true is not widely known among the general population, but has been confined to a narrow circle of scholars and researchers in the history of science and Hermeticism.

As we indicated in the previous chapters, the Hermetic metaphors of the Tower of Babel Moment and of the "Topological Metaphor" of the physical medium concerned themselves primarily with three things: (1)physics, (2) life, or biology, and (3) the mediating principle between the two: consciousness, expressed in the image of androgyny. These themes are detectable as Hermetic influences within modern science, and a brief overview of them is necessary before we can appreciate the alchemical basis for the quest of man to create and alter life itself.

A. A BRIEF REVIEW OF THE HERMETIC BASIS OF MODERN PHYSICS

Before pressing into the alchemical basis of the goals of modern genetic engineering, a brief review of the influence of hermeticism on the rise of modern science will help to illustrate the vast though largely hidden influence of occult doctrines on science. We need only consider the influence of Hermeticism on Copernicus, Kepler, Newton, and Leibniz.

1. Hermeticism in Copernicus and Kepler

Frances A. Yates was a scholar of mediaeval and Renaissance culture, and exposed some of these roots in a critically important work, *Giordano Bruno and the Hermetic Tradition*. As Yates points out, the influence of the ancient theology, the *prisca theologia*, on Copernicus was acknowledged by the astronomer himself, and in his revolutionary exposition of the heliocentric universe:

> The *De revolutionibus orbium caelestium* of Nicholas Copernicus was written between 1507 and 1530, and published in 1543. It was not by magic that Copernicus reached his epoch-making hypothesis of the revolution of the earth around the sun, but by a great achievement in pure mathematical calculation. He introduces his discovery to the reader as a kind of act of contemplation of the world as a revelation of God, or as what many philosophers have called the visible god. It is, in short, in the atmosphere of the religion of the world that the Copernican revolution is introduced. Nor does Copernicus fail to adduce the authority of *prisci theologia* (though he does not actually use this expression), amongst them Pythagoras' and Philolaus to support the hypothesis of earth-movement.... Copernicus is not living within the world-view of Thomas Aquinas but within that of the new Neoplatonism, of the *prisci theologia* with Hermes Trismegistus at their head.... One can say, either that the intense emphasis on the sun in this new world-viw was the emotional driving force which induced Copernicus to undertake his mathematical calculations on the hypothesis that the sun is indeed at the centre of the planetary system; or that he wished to make his discovery acceptable by presenting it within the framework of this new attitude. perhaps both explanations would be true, or some of each.[2]

In other words, while Copernicus' mathematics was not hermetically inspired,[3] the overall inspiration for the hypothesis may have been, since Copernicus himself cites the favorite sources of the Renaissance magicians, namely, the Pythagoreans, the Neoplatonists, and Hermes Trismegistus himself.

The other great "hermetic scientist," Kepler, was even more heavily influenced by Hermeticism, and yet, like Copernicus, was able to successfully dissociate his mathematics from Hermetic number-mysticism:

> The mighty mathematician who discovered the elliptical orbits of the planets had, in his general outlook, by no means emerged from Renaissance influences. His heliocentricity had a mystical background; his great discovery about the planetary orbits was ecstatically welcomed by him as a confirmation of the music of the spheres; and there are survivals of animism in his theories. Nevertheless, Kepler had an absolutely clear perception of the basic difference between genuine mathematics, based on quantitative measurement, and the "Pythagorean" or "Hermetic" mystical approach to number.[4]

What was happening, in other words, was that these scientists remained influenced by hermetic doctrines and beliefs about the nature of the universe, but were successful in translating those doctrines into a new kind of mathematics liberated from centuries of accreted numerological mysticism. This is particularly the case with respect to the two great mathematicians, each of whom independently invented the differential and integral calculus, and both of whom were heavily influenced by hermetic and alchemical doctrine: Newton and Leibniz.

2. In Newton

Sir Isaac Newton is undoubtedly one of the world's best known, if not *the* best known, scientist. What most do not know about Newton, however, was that "scientist" seems to be at best an honorific label extended to him by scientists for the greatest of his ideas, his theory of gravity, but it was not, perhaps, a label he would have comfortably worn himself. With his theory of gravity, it was no longer possible to doubt the correctness of the Copernican-Kepleran heliocentric theory.[5] The reason for this uncomfortable label was simply that Newton was *not* a scientist; he was an alchemist, a magician:

> On his death, 169 books on alchemy were found in his personal library - making up one-third of his collection. In fact, it transpires from all his writings that his main esoteric preoccupation was the quest for the philosopher's stone, and he was particularly fascinated by the work of the French alchemist Nicolas Flamel (c. 1330-1418).
>
> Most of Newton's alchemical papers - of which he produced a vast number, over a million words - collected by Keynes and others,

are now in Jerusalem, in the Jewish National Library. As befits the work of a genius with a need to be secretive, they are written in elaborate codes, and many of them have yet to be deciphered.[6]

The mention of the economist John Maynard Keynes brings up the fact that he, and other scientists, began the process of begrudgingly admitting that one of the world's great scientific minds was heavily steeped in activities and beliefs that can only be classified as magical, hermetic, and alchemical, and which perforce were hardly "scientific." Viewed a certain way, in other words, modern science may be viewed as "Alchemy, upgraded."

For example, Richard Westfall, a professor of the History of Science at the University of Indiana, put the problem of Newton's esoteric interests thusly in his 1972 biography of Newton: it had to be admitted, observed Westfall, that there were present "in Newton's mind modes of thought long deemed antithetical to the modern scientific mind."[7]

Keynes himself put the point with much more eloquence in an address given to the Royal Society in 1946:

> Newton was not the first of the Age of Reason. He was the last of the magicians, the last of the Babylonians and Sumerians, the last great mind which looked out on the visible and intellectual world with the same eyes as those who began to build our intellectual world rather less than 10,000 years ago... Why do I call him a magician? Because he looked on the whole universe and all that is in it *as a riddle*, as a secret which could be read by applying pure thought to certain evidence, certain mystic clues which God had laid about the world to allow a sort of philosopher's treasure hun to the esoteric brotherhood. He believed that these clues were to be found partly in the evidence of the heavens and in the constitution of elements (and that is what gives the false suggestion of his being an experimental natural philosophers), but also partly in certain papers and traditions handed down by the brethren in an unbroken chain back to the original cryptic revelation in Babylon.[8]

Newton, in other words, was not a modern scientist in one important respect, namely, that in addition to the observations of nature, Newton incorporated into the "dataset to be interpreted" a vast collection of manuscripts and traditions that in his view *may* have incorporated the legacy of a lost civilization and its high science. He was, as such, "a great believer that the earliest civilizations, such as Egypt, knew more than people in his own day - that they possessed the *prisca sapientia*, or 'ancient wisdom.'"[9]

Part of that *prisca sapientia* in Newton's case, came from the actual *Hermetica* themselves, the texts we examined briefly in chapter two in connection with the "Topological Metaphor" of the physical medium. This text, as it turns out, is the ultimate source for Newton's ideas about gravity:

> It is not simply a matter of Newton hitting on the physical laws of nature by drawing analogies with the Hermetic principles. He *applied* those principles to physical systems. For example, the big resistance to his explanation of gravity was that many considered it to be too 'occult.' His notion of gravity as a force that acts across space, at a distance, and does so in the way it does purely as a consequence of the nature of the universe, was drawn straight from the magical laws of sympathy and attraction as expounded in the Hermetica. (Newton put it more succinctly, declaring 'Gravity is God.') The law of gravity invokes principles relating to forces that act between the Earth and heavenly bodies that feature - in very different language of course - in *Asclepius*, the same work that inspired Copernicus.[10]

Newton himself gives a glimpse of this "scientific hermeticism" at work at the very end of his *Principia*. There, in language meant to evoke the Neoplatonic and hermetic doctrine of the World Soul or World Spirit, he outlines the course for future science: to discover the laws of that Spirit's operation, and thereby, to gain what all magicians strive for: mastery over it:

> And now we might add something concerning a certain most subtle Spirit which pervades and lies hid in all gross bodies; by the force and action of which Spirit the particles of bodies mutually attract one another at near distances, and cohere, if contiguous; and electric bodies operate to greater distances, as well repelling as attracting the neighbouring corpuscles; and light is emitted, reflected, refracted, inflected, and heats bodies; and all sensation is excited, and the members of animal bodies move at the command of the will, namely, by the vibrations of this Spirit.... But these are things that cannot be explained in few words, nor are we furnished with that sufficiency of experiments which is required to an accurate determination and demonstration of the laws by which this electric and elastic Spirit operates.[11]

In other words, if Newton's *Principia*, in which he elaborated his theory of gravity, "had never been written, our moden technological world would not exist. But without the Hermetica, Newton would never have written the *Principia*. Emphatically, Newton did not make his great scientific discoveries *despite* his esoteric beliefs, but *because* of them."[12]

3. In Leibniz
a. Leibniz's Characteristica Universalis and the Quest for a Universal Formal Language

If the hermetic and alchemical influences are strongly evident in Newton, they are even more so in the other great genius of the time, Gottfried Leibniz who, along with Newton though quite independently of him, is credited with inventing the differential and integral calculus, and whose notation conventions for the calculus are in use to this day.

Like Newton, Leibniz was an alchemist, but additionally, Leibniz's late works display a close familiarity with the writings and doctrines of Rosicrucianism, and his first important work, *Dissertation on the Art of Combination* is nothing but a discourse on the art of memory, in which he acknowledges his indebtedness to the great Renaissance magus and practitioner of the art of memory, Giordano Bruno.[13]

Leibniz deliberately conceived his invention of the calculus, however, to be but the first step in a much more ambitious project, the invention of a *characteristica universalis*, a universal formally explicit language able to handle both quantized, and non-quantized, concepts and thinking:

... (Even) if Leibniz was wary of shouting it from the rooftops, his works quite clearly owe a major debt to the Renaissance occult philosophy. Even Leibniz's system of calculus evolved from this tradition. it developed from his quest to reduce everyything, not just scientific principles and laws *but also religious and ethical questions, to a common symbolic language: a universal calculus.* Builing on the art of memory, both the classical and 'occult' versions, in order to establish a language of symbols or *characteristica universalis*, Leibniz envisaged a set of mages to which all the fundamentals of knowledge could be reduced. This naturally necessitated the cataloguing and codification of all that was known, a growing eighteenth-century preoccupation. By manipulating and setting the symbols in different relationships, he believed that new discoveries could be made.

He specifically likened such a system to Egyptian hieroglyphs, which along with Bruno, he believed were used in a similar way.... Leibniz even described his *characteristica universalis* as 'true Cabala'- hardly the words of a modern-style rationalist.

Eventually Leibniz came to realize that the best tools for the job were mathematical symbols. This realization then led to the development of his version of infinitesimal calculus, which he intended to be a first step towards the universal calculus. [14]

To put it differently, what Leibniz was aiming for was a formally explicit analogical calculus, able to handle the modeling of pure forms across several disciplines in various combinations, allowing one to "calculate" by means of "pure concepts."[15] We believe that Leibniz was aiming for a universal formal language that could describe what we are calling the "Topological Metaphor" of the physical medium, and that could describe all its differentiations.

B. The Alchemical Basis of Modern Genetic Engineering

It is when one turns to the hermetic and alchemical influences upon the other science with which we are preoccupied - biology - that the case is made more difficult. Here, we are confronted with biologists whose libraries and writings and technical arsenals are not filled - as with Newton, Leibniz, Copernicus, or Kepler - with alchemical books and references. Thus, to discover an alchemical influence, one must look for detailed parallels not only in basic *goals*, but also in the *analogous techniques* between alchemy on the one hand, and biology on the other.

We are, however, in a fortunate position in this respect, for alchemy and modern genetic engineering share an important, and one might also say, "Promethean Ambition,"[16] to manipulate, engineer, and eventually, even create life itself, a goal that alchemy specifically gave shape and form in the "homunculus," an artificially created man, and here, once again, we are chin-to-chin with that disconcerting image of androgyny.

1. The Promethean Alchemist's Ambition:
The Creation and Manipulation of Life in the Homunculus
a. The Dream of Reanimation and Virtual Immortality

We begin in an odd place, and with a virtually unknown man, the Venetian metallurgist and cannon-maker Vannoccio Biringuccio (1480-ca. 1539), who had little faith in alchemical doctrines, and yet, noted that alchemy had indeed led to many useful discoveries.[17] Nonetheless, Biringuccio disputes the basic tenets of alchemy and the claims of the alchemists themselves as being impious:

"What greater folly could men commit than to waste their time in following the other arts and sciences and to fail to study and learn this art (alchemy) which is so useful and so worthy, nay divine and supernatural?" Here we see precisely the competition between alchemy and the other arts exposed in clear language. The problem with alchemy is that its claims make it not only the Queen of the Arts, but effectively the only "real" art, since alchemy along can genuinely master nature. In reply to this claim, Biringuccio responds again with the charge of impiety - if the alchemists really had an elixir that could transmute whatever metal they desire into gold, they could say :that they hold prisoner in a bottle that God which is the creator of all these things." But even this is not the greatest of their claims.

Beyond transmuting metals, *Biringuccio's alchemists also maintain that they can convert bread, herbs, and fruit into flesh by means of artificial digestion in a flask.* They can even make carbonized wood green again, whereon it will bud and produce more wood. It appears that Biringuccio had already encountered the alchemical project of artificial life.... He develops this topic in the following fashion:
"With this and many other reasons they wish to make you believe that even outside a woman's body it is possible to generate and form a man or any other animal with flesh, bones, and sinews, and to animate him with a spirit and every other attribute that he requires. And in like manner *they say it is possible by art to cause trees and grasses to be born without their natural seeds*, and to give fruits separated from trees the form and color, odor and flavor of true natural fruits."[18]

The assertions of the alchemists of his day pointed out by Biringuccio raise certain important questions, questions that are obvious to us today, who are used to the production of "life outside a woman's body" and all other sorts of bizarre pursuits of modern scientists, not the least of which is the curious resemblance of Biringuccio's alchemists' statements that "it is possible by art to cause trees and grasses to be born without their natural seeds" to, as we shall see later on in the next chapter, agribusiness's "seedless seeds."

The problem is precisely that of anachronism: why, at a time when our modern genetic and medical techniques were totally unknown, were alchemists claiming such things? We are faced with two possibilities:

1) The alchemists were making wild and fantastic claims with absolutely no basis in truth or previous history; or,
2) The alchemists had actually managed, somehow, to do such

things from time to time, and this raises the possibility that they were preserving a lost high knowledge from antiquity in making these claims, a knowledge and goal perhaps eventually recovered by modern genetics. Indeed, how else would one account for the curious statements we shall encounter a little further on, that life can be made in a "vessel" or "flask" if it did not, in some way, reflect experience with an actual technology?

Strangely, the alchemical idea of virtual immortality, of the rejuvenation and resuscitation of that which is dead finds an exact parallel in modern science.

Modern genetic techniques have allowed researchers to do precisely what Biringuccio's alchemists claimed: growing body parts "outside" the natural environment - test tube babies - and to envision a whole new field of therapy: regenerative medicine, with the possibility "for humans to regenerate a damaged body part the way starfish and salamanders can..."[19] Bladder, skin, trachea, blood vessels and cartilage substitutes have already been grown in laboratories and successfully used in therapies.[20] Eventually, it will be possible to grow kidneys, hearts, and other organs according to a specific patient's genetic makeup, both decreasing the dangers of transplants that are rejected, and the current situation where patients oftentimes wait for organs to become available for transplant. Beyond this, burns and spinal injuries are another area in which modern technique appears to be catching up to the alchemical claims of four centuries ago.[21]

b. The Androgyne at the End of the Age:
The Alchemical Apocalypse and Final Transformation of Matter

Birringuccio also reported that the alchemists of his day claimed dominance "not only over all the things of this world, but of the next."[22] This apocalyptic and eschatological goal found its strongest expression in the fourth century Greco-Egyptian alchemist and Gnostic, Zosimus of Panopolis, a follower - once again - of Hermes Trismegistus and the body of writings attributed to him, the *Hermetica*.[23]

The material world, according to Hermes, is animate and ensouled, but it was corrupted by the Fall. Zosimos adopts this idea wholeheartedly and endues the alchemist with a strong sense of religious purpose - liberating the world from sin. He should do this literally by purging matter of its dark and heavy attributes. By a process involving distillation, purification of residues, and other operations, Zosimus and his contemporaries hoped to remove the impurity of matter and to make it pneumatic, thus "resurrecting" the material world.[24]

In other words, alchemy had its own eschatology in the final transformation of matter, a transformation the pinnacle of which was the final alchemical transformation of mankind itself.

It is thus with Zosimus that we encounter the first glimpse of the alchemical manipulation of life itself, and of its claim to be able to create an artificial humanoid life form, the homunculus, and the imagery, again, is eerily modern, recalling yet again the anachronism of alchemy and its all-too-prescient foreknowledge of the subsequent course of biological science:

> The vision of Zosimus begins... with a priest contained within an alchemical vessel who is being converted from gross matter into subtle pneuma. It is likely that the actual process involved is distillation, since the term used for the vessel, *phiale*, is employed elsewhere by Zosimus to mean a part of a still. The image of a man inside a flask already conjures up images of artificial life.
>
> This interpretation may seem at first to be confirmed when Zosimus says that the priest becomes an *anthroparion* - a little man or homunculus - upon mutilating himself. It is true that this image opened up a major iconographical tradition in alchemy - the Middle Ages saw the creation of numerous illustrations of men, women, and animals in alchemical bottles.... Indeed, the theme became fused with the biological concept of the alchemical process as a form of Holy Matrimony - a *heiros gamos* - where chemical substances were thought to combine by a process like copulation and to give birth ultimately to a glorious substance called the philosophers' stone. Hence one commonly finds illustrations of kings and queens sealed up in flasks copulating and giving birth. At the same time, Zosimos's theme of ritual purification and chastisement lent itself to the notion that the substances in the flask must be punished, killed, and reborn in a glorious, regenerate state. Needless to say, this conformed nicely to the Christian myth of death and rebirth, so that one frequently finds the alchemical couple dying and being regenerated, sometimes the couple even becomes a hermaphrodite, which is usually killed and reborn.[25]

The "hermaphrodite at the end of time", however, is less a Christian image than it is the image of the ancient theology and its Topological Metaphor of the medium, with mankind finally returning to its assumed original "alche-mosexual" androgyny.

This alchemical eschatology and apocalyptic androgyny brings us to the most famous alchemist of them all, and to his unique version of the alchemical apocalypse theater:

c. Paracelsus

By the time of Paracelsus(1493-1541), whose full name, Philippus Aureolus Theophrastus Bombastus von Hohenheim, gives some measure of the man, the creation of an artificial man, the homunculus, was considered to be, along with the confection of the Philosophers' Stone itself, the crowning achievement of alchemical art.[26]

With Paracelsus, the androgynous homunculus as a major theme of alchemical art comes home, and with a vengeance in the man himself, for recently his remains were recently exhumed, and researchers came to an astonishing conclusion concerning his sex, for researchers:

> found that his pelvis was extraordinarily wide, indicating a high probability that he was suffering from some form of intersexuality. Since his extremities betray none of the lengthening associated with eunuchs who have undergone prepudescent castration, the forensic specialists suggest that Paracelsus was either a genetic male afflicted with pseudohermaphroditism or a genetic female suffering from adrenogenital syndrome. In the latter case, the clitoris enlarges during fetal development to assume the appearance of a penis, and the labia can fuse together to form a structure like an empty scrotum - hence the early reports of Paracelsus's castration might be based on eyewitness accounts of his genitalia. At any rate, we are left with the remarkable possibility that the gender of Paracelsus may have been capable of description as either female or male.[27]

Paracelsus' possible "intersexuality" places his alleged comments and writings about the creation of an artificial homunculus into an intriguing context.

In 1572 the physician and Paracelsan alchemist Adam von Bodenstein "published a work supposedly written in 1537 by Paracelsus"[28] called the *De natura rerum*. This treatise raises many of the same distinctions, both ethically and philosophically, posed by the modern techniques of genetic engineering, but in an alchemical context:

The generation of all natural things is of two sorts, as (there is) one that happens by means of nature alone without any art, (while) the other happens by means of art - namely alchemy. In general, however, one could say that all things are born from the earth by means of putrefaction. For putrefaction is the highest step, and the first beginning of generation, and putrefaction takes its origin and beginning from a moist warmth. For the continual moist warmth brings about putrefaction and transmutes all natural things form their first form and essence, as also their powers and virtues. For just as the putrefaction in the stomach turns all food to dung and transmutes it, so also the putrefaction that occurs outside the stomach in a glass (i.e., a flask) transmutes all things from one form into another.[29]

Later in the work, reference is made to the "fire of the Day of Judgment,"[30] which in the context implies the high heat of alchemical transformations. The *De natura rerum* also makes it clear that a kind of alchemical virgin birth or parthenogenesis is possible through alchemical techniques: "You must also know that men too may be born without natural fathers and mothers. That is, they are not born from the female body in natural fashion as other children are born, but a man may be born and raised by means of art and by the skill of an experienced spagyrist..."[31] The idea of transforming and transmuting "putrefaction" in a flask or alchemical vessel has a long history, one that again raises the image of androgyny.

For example, Arab alchemists record the story of a king who, wishing a male heir, consulted an alchemist, who assures him that all that is needed was some of the king's sperm, which would then be "kept in a vessel" to which techniques would be applied, resulting in a male son.[32] This makes its appearance in the *De natura rerum* in the form of the homunculus, which is - in yet another return to the image of the primordial masculine androgyny - "the distilled essence of masculinity...(because) of its freedom from the gross materiality of the female."[33]

This returns us to a theme which we encountered in the first chapter with the Mayans, i.e., to the idea that within the presumed original state of androgyny, humanity, by dint of that state, somehow possessed greater knowledge. The idea returns in the *De natura rerum* in connection with the homunculus, which, because it is born by "art," has "art" innate to it, and need not "learn it from anyone."[34] Because it is created by alchemical art, "in its mature state it has an automatic and intimate acquaintance *with* the arts, and consequently knows 'all secret and hidden things.'"[35] The androgynous homunculus, in other words, is the alchemical reversal of the Fall, of the

Tower of Babel Moment, and thus the creation and manipulation of life is a necessary stage toward the apocalyptic fulfillment of that goal. With this in mind, we must now turn to a closer consideration of the actual techniques - both alchemical and scientific - of the manipulation and engineering of life, first, with a closer look at what Paracelsus has to say on the matter.

2. Paracelsus on the "Techniques" of Engineering the Homunculus

Paracelsus, like most alchemists and hermeticists, believed that mankind was a microcosm, only in Paracelsus' case, mankind was so because he was made of the "dust of the ground," and not created *ex nihilo*. As such, mankind contained within himself the powers of creation,[36] among these were, of course, that mankind had a "power of androgyny." In Paracelsus' case, however, this was recounted in a rather graphic and descriptive language outlining the "technique" or "technology" involved in the creation of the homunculus. For Paracelsus, man was *already* chimerical, possessed of a spiritual soul and an *animal* body, and thus was already to a certain extent already embarked upon the first step of the ladder of alchemical ascent:

> Now the animal body of man exists independent of the soul, and it produces a defective, soulless sperm when one is possessed by it. It is from this defective, soulless sperm, Paracelsus now tells us, that homunculi and monsters are produced: therefore they have no soul.
>
> But this can happen in different ways. First, as soon as a man experiences lust, sperm is generated within him. He has a choice at that point; he may either act on his lust and let the semen pass out, or keep it within, where it will putrefy internally. If he should allow the semen to pass out of his body, it will proceed to generate as soon as it lands on a *Digestif* - that is, a warm moist subject that can act as an incubator. This "polluted sperm" must produce a monster or homunculus when it is "digested."[37]

This result Paracelsus refers to a "Sodomitic birth," and for him the alchemical possibility of the production of a homunculus thereby constitutes the real reason for the Church's sanctions against it.[38]

Yet, in the *De natura rerum*, this act is turned into a virtue. There, one may incubate a flask at the proper temperature, and then "isolate the male seed from the female and so produce a transparent, almost bodiless homunculus. *In this fashion, human art can generate a being unimpeded by the materiality of normal female birth, hence surpassing the artifice of nature itself.*"[39]

There are a number of points that must now be considered, and the first of these is the authorship of the *De natura rerum* itself, which most scholars doubt to be by Paracelsus at all, since in his genuine works, the great alchemist is utterly disposed against "sodomitic births." Here we are confronted with the possibility that Paracelsus might actually have authored the *De natura rerum*, choosing the safe course in his other writings, but revealing his real thoughts in a work published posthumously.

But whether the *De natura rerum* represents Paracelsus' real views or not, it certainly *does* represent the views of alchemy. And here one must note two very crucial things:

1) The homunculus is essentially viewed as an androgyne, the product solely of male seed, and as such is viewed as a perfection of nature; and,

2) It is produced by technical means that, in its general descriptions at least, resembles a *tissue culture* which is incubated at the proper temperature in a flask. We have, in other words, an alchemical version of a test tube baby.

This production was viewed as "the crowning pinnacle of human art."[40] For these alchemists, "male parthenogenesis" was the highest goal, for it was understood by them to be a method whereby to escape the material world, and to reascend the ladder of topological descent to the higher androgynous humanity. [41] It was, in short, an apocalyptic, eschatological goal. But why, once again, is this "androgyny" persistently and consistently viewed by so many, in masculine terms, as a *male* rather than a *female* androgyny? The answer to that question must await a later chapter.

C. CONCLUSIONS THUS FAR

For now, we must now pause briefly to review what we have found, for the alchemical nature of the goals of modern science and genetic engineering may not readily be appreciated without such a review. Because the image of androgyny functioned as a symbol for the blending of all manner of characteristics and not just sexual ones, it becomes a symbol for the re-ascent up the ladder of the metaphor of the descent of man, implying four fusions: (1) animal and man, (2) vegetable, or plant and man, and finally (3) mineral, or machine, and man; and finally, (4) androgynous man in the proper sense.

To put it differently, the alchemical goal is to transform man the microcosm from a theory to reality, and the way to do this is to create the fusions by "art," i.e., by a technique and technology, to fill the space of creation with man, quite literally by "splicing" or merging him with the animal, vegetable, and mineral kingdoms. Let us recall in this regard two important points that were observed about the claims of alchemy as recorded by the Venetian cannonist and metallurgist Vannoccio Biringuccio:

1) "Bread, herbs, and fruit," that is to say, the plant or "vegetable" kingdom, could be "converted" into flesh *"by means of artificial digestion in a flask"*(p. 105); and,
2) It is also possible by the alchemical art to "cause trees and grasses to be born without their natural seeds,"(p. 108).

As we shall discover in the next chapter, the second point, the generation of seedless seeds, has in fact been a supreme goal of the agribusiness industry. What concerns us here is the first point., for the alchemical art implied the fusion of plant and human characteristics, implying the ability to fuse human and other animals. Indeed, given the doctrine of man the microcosm, and the goal of transforming that doctrine into a reality by means of alchemical techniques, it should not surprise us that such "enhancements" or "perfections of nature" are indeed taking place within modern genetic engineering, with the creation of "manimals" or what we prefer to call alchemomanimal man. To this subject we now turn.

D. Chimeras:
Alchemanimal Man, the Law, and Social Engineering

We have come to the point - the alchemical fusion of man and animal - where we begin our examination of the alchemical "re-ascent" back up the Tower to the original and primary event. We therefore state here the thesis that shall guide us throughout the remaining chapters of this section of the book, and on into the next section: *all technologies and techniques aiming for the transformation of mankind, or for re-engineering him, is alchemy, and also constitute the alchemy of social engineering.* In other words, by fulfilling the goals of the creations of "androgynous fusions" and the transformations of consciousness resulting therefrom, the techniques of modern science are revealed to be nothing but the perfections of alchemical "pseudo-science," but the goals remain the same. It is also important to note that we consider here but a few examples out of thousands that can be researched on the internet, and the potentials and implications they raise.

Consider the first implication of the blending of animal and man in the following cases reported by *National Geographic News:*

Chinese scientists at the Shanghai Second Medical University in 2003 successfully fused human cells with rabbit eggs. The embryos were reportedly the first human-animal chimeras successfully created. They were allowed to develop for several days in a laboratory dish before the scientists destroyed the embryos *to harvest their stem cells.*

In Minnesota last year researchers at the Mayo Clinic created pigs with human blood flowing through their bodies.

And at Stanford University in California an experiment might be done later this year to create mice with human brains.

Scientists feel that, the more humanlike the animal, the better *research model it makes for testing drugs or possibly growing "spare parts," such as livers, to transplant into humans.*[42]

At this juncture, the *National Geographic* article asks the significant questions, ones to which we shall return a little later: "At what point would it be considered human? And what rights, *if any*, should it have?"[43] Note the carefully couched but deliberate moral ambiguity that is being planted here, for if such a creature were to be defined as human, then perforce, it should be possessed of the full panoply of human rights. The ground of jurisprudence, in other words, is being carefully prepared for a transformation of consciousness, for a dramatic program of alchemical social engineering, for a redefinition of what it means to be human.

But more of this in a moment. For now, we remain focused on the already-accomplished chimerical creations and speculations of the alchemical genetic engineers. Consider yet another implication of these techniques:

> For example, an experiment that would raise concerns, (David Magnus) said, is genetically engineering mice to produce human sperm and eggs, then doing in vitro fertilization to produce a child whose parents are a pair of mice.[44]

Would such an offspring be considered human because its genetic composition was predominantly human, or somehow less-than-human because of the possible admixture of a certain amount of "mouse" genes in its genome? What of the case where Standford's Irv Weissman was considering genetically engineering mice to have 100 percent human brains? "This would be done, he said, by injecting human neurons into the brains of embryonic mice. Before being born, the mice would be killed and dissected to see if the architecture of a human brain had formed. If it did, he'd look for traces of human cognitive behavior."[45]

But if such "alchemomice" or "manimice" were to have "traces of human cognitive behavior," does this not raise the possibility that they would have a human sense of "self" or something approaching it, and have human emotions and feelings, or something approaching them? And if this be the case, should they not have some protection or recognition of rights, such as to prevent their wanton destruction in the name of science run amok? But conversely, would it be safe or wise to allow such creatures to propagate, given that the environmental effects - should these "manimice" get into the general population[46] - are unknown? Who wants to combat mice with human intelligence in their house in some kind of bizarre "Rodent Wars"?

On and on we could go. We have already noted the creation of mice with human brains and pigs with human blood, but it does not stop there. There have now been "sheep whose livers and hearts are largely human," (note the ambiguity, *largely* human), engineered in Nevada, with livers making all the compounds of normal human livers,[47] reports of human embryos, and even of human-animal chimerical embryos, created through cloning,[48] and mice with human immune systems.[49]

And remember those Minnesota pigs at the Mayo Clinic with human blood? Well, it seems that in the normal course of evolutionary mutation, something happened, for it's no longer a case, of "just pig blood cells being swept along with human blood cells' some of the cells themselves have merged, creating hybrids."[50] This is an important point, with profound legal implications that we will explore again in the next chapter.

At this point, we must pause to set up the necessary context within ancient lore and mythology to understand the speculative chimerical creations now being seriously entertained in genetic science. Joseph has written many previous books touching upon the prospect that humanity itself was genetically engineered in ancient times as a slave race to "the gods," who, as it turns out, were probably members of the human genus, that is to say, they were our "genetic cousins" to begin with.[51] Robert Streiffer, professor of philosophy and bioethics at the University of Wisconsin, has raised the specter of the creation of "a human-chimpanzee chimera endowed with speech and an enhanced potential to learn - what some have called a 'humanzee.'"[52] In other words, a chimerical slave race - humanzees - to serve a race of "gods" in the performance of "menial jobs or dangerous jobs,"[53] "gods" who are its genetic cousins! The *Planet of the Apes* is already a potential reality.

And what of the mice with human brains that might show "traces of human cognitive behavior"? The danger is real:

The potential power of chimeras as research tools became clear about a decade ago in a series of dramatic experiments by Evan Balaban, now at McGill University in Montreal. Balaban took small sections of brain from developing quails and transplanted them into the developing brains of chickens.

The resulting chickens exhibited vocal trills and head bobs unique to quails, proving that the transplanted parts of the brain contained the neural circuitry for quail calls. It also offered astonishing proof that complex behaviors could be transferred across species.[54]

We raise these examples not only to show that the alchemical and hermetic agenda of creating "androgynous fusions" at the level of "animal" man are not only alive and well, and being actively pursued, but because there is a second way in which these fusions are functioning alchemically, and that is, at the level of mankind's *social environment and consciousness*. And the transformation of consciousness is, after all, one of the principal goals of alchemy.

But how does the creation of "alchemomanimals" transform man's consciousness? We believe the answer to this question is quite simple, and easily illustrated by its immediate social consequence: *law and jurisprudence*. How *much* human genetic material can be in a "manimal" before it is more human than animal? And what, if any, rights in law should it (or he, or she) have? Does a creature with 80 percent human and 20 percent animal DNA have less rights than a "full" human because it (or he, or she), is not fully human? If genetics is forcing on us the requirement to think in *gradations* of species within a creature, then genetics is making possible, indeed is *forcing* the possibilities that once led the U.S. Supreme Court in the Dred Scot decision to define a black human being as "3/5 of a person." What was ridiculous then, is now being raised in a much more direct and blunt fashion by genetic-alchemical techniques: will genetic chimeras result in similar gradations or "defined rights" in law, and therefore, in mankind's social environment and culture? We are reminded of Percy Bysshe Shelley's words in *Frankenstein* (and yes, we said *Percy* Shelley, not *Mary* Shelley, a point we shall explore in chapter nine): "The dissecting room and the slaughter-house furnished many of my materials; and often did my human nature turn with loathing from my occupation, whilst, still urged on by an eagerness which perpetually increased, I brought my work near to a conclusion."[55]

Nor is it only the creation of chimeras that raises these issues, but, as Shelley strongly suggested in *Frankenstein*, also the *manner of their generation,* and again, it was not modern science, but mediaeval alchemy which first raised the debates:

The homunculus, or miniature human created in an alchemical flask, was a topic of discussion already among the medieval Arabs. Could one use this form of generation to alter the sexuality of the child? Why not make a being of extraordinary intelligence, *with powers denied to the offspring of normal sexual generation?* Was it permissible to use the bodily fluids of the homunculus as a means of curing dangerous disesaes? Have we not heard all of these questions discussed recently in the controversy surrounding the artificial selection of gender, the prenatal modification of biological traits, and the use of fetal tissue for medical purposes?[56]

Was an alchemically created human, engendered without normal human sexual generation, even human at all?

We saw earlier in this chapter that "parthenogenesis" via a male's sperm was an actual apocalyptic goal of alchemy. Thus, the creation of humans with only a father - a kind of male parthenogenesis which, as we saw, *was an explicit goal* of alchemy - is at least potentially possible, and raises the issue of the homunculus: are such creations to be considered fully human, or not? And if not, why define *humanity* or "human nature" to be dependent simply on the manner of one's generation, since the great Yahwisms themselves acknowledge that Adam and Eve, though brought into existence without the normal means of human generation, were nonetheless fully human?

Questions such as these illustrate the high degree to which genetic engineering is *also social engineering, and therefore, is also the alchemical transformation of human consciousness and society.*

There is, however, another implication. Under U.S. patent law, an invention must fulfill four requirements in order to be patentable as *intellectual property:*

1) It must be original, i.e., it must not have been published or patented previously nor be too similar to a previous invention;

2) It must not be obvious, that is to say, you cannot patent a rock wrapped in a sock and call it a "non-scuff" door stopper;

3) It must have a clear and demonstrable *function*, which, in the case of chimerical life, as we have seen, could include specific research purposes, such as the study of immune system disorders in chimerical creatures with human immune systems, or human cognitive or mental disorders in creatures with predominantly human neural structures;

4) It must be enabling, in other words, the patent should function like a recipe, with clear enough descriptions of the technologies and techniques to allow anyone to reproduce it.[57]

Under these criteria, an animal-human chimera, even if, say, 99 percent human and only 1 percent animal (or vice versa), would be a *patentable object and intellectual property*....

....shades of the ancient Mesopotamian and Meso-American myths of the engineering of mankind as a slave race to the gods.

If this seems farfetched, or the remotest thing from the minds of the power brokers behind the genetic magicians, then brace yourself, for we are now going to ascend to the next level of "androgynous symbolism," the alchemical wedding of the human and the vegetable, and the alchemical "seedless seeds" of agribusiness and the banksters, and the principle of "substantial equivalence."

ENDNOTES

1   William R. Newman, *Promethean Ambitions: Alchemy and the Quest to Perfect Nature* (The University of Chicago Press, 2004), p. 1.
2   Frances A. Yates, *Giordano Bruno and the Hermetic Tradition*, pp. 153-154.
3   Ibid., p. 155.
4   Ibid., p. 440.
5   See the discussion in Lynn Picknett and Clive Prince, *The Forbidden Universe: The Occult Origins of Science and the Search for the Mind of God*, pp. 163-165.
6   Picknett and Prince, op cit., p. 167. This fact does raise the interesting question of why Newton's alchemical papers should have ended up there, and why they did so.
7   Ibid., p. 166, citing Richard S. Westfall, "Newton and the Hermetic Tradition," in Debus (ed.), *Science, Medicine and Society in the Renaissance*, Vol. II, pp. 185-186.
8   John Maynard Keynes, "Newton the Man," in *The Royal Society*, New Tercentenary Celebrations (1947), p. 29, cited in Giorgio de Santillana and Hertha con Dechend, *Hamlet's Mill: An Essay on Myth and the Frame of Time* (Boston, 1969: Gambit Incorporated), p. 9.
9   Picknett and Prince, *The Forbidden Universe,* p. 167.
10 Ibid., pp. 170-171, emphasis in the original.
11 Isaac Newton, *The Principia*, Trans. Andrew Motte, *Great Minds Series* (Prometheus Books, 1995), p. 443.
12 Picknett and Prince, *The Forbidden Universe.*, p. 172.
13 Ibid., p. 129.
14 Ibid., p. 131, emphasis added.
15 Volker Peckhaus, "Calculus Raiocinator vs. Characteristica Universalis? The Two Traditions in Logic, Revisited, p. 6.
16 The phrase "Promethean Ambition" is the title of a vitally important book by William R. Newman, *Promethean Ambitions: Alchemy and the Quest to Perfect Nature*, which we will follow closely in this section. In our opinion this is the best single volume study of the alchemical agenda to perfect nature.
17 William R. Newman, *Promethean Ambitions: Alchemy and the Quest to Perfect Nature* (The University of Chicago Press, 2004), p. 127.
18 Ibid., pp. 129-130, emphasis added, citing Vannoccio Biringuccio, *De la Pirotechnia.* 1540. Facsimile, ed. Adriano Carugo (Mlian Polifolio, 1977). Translated into English as *Pirotechnia,* Trans. Cyril Stanley Smith and Martha Gnudi (Cambridge, MA: MIT Press, 1942), 85(facs.)/43.
19 Anthony Atala, "Regenerative Medicine's Promising Future," CNN, July 10, 2011, www.cnn.com/2011/OPINION/07/10/atala.grow.kidney/ index.html, p. 1.
20 Ibid., p. 2.
21 Ibid.
22 William R. Newman, *Promethean Ambitions*, p. 130.
23 Ibid., p. 171.
24 Ibid., pp. 171-172.
25 Ibid., p. 173.
26 Ibid., p. 199.
27 Ibid., pp. 196-197.
28 Ibid., p. 199.
29 Ibid., p. 200, citing Pseudo-Paracelsus, *De natura rerum*, in Sudhoff, 11:312.
30 Ibid. p. 201.
31 Ibid., pp. 201-202, citing Pseudo-Paracelsus, *De natura rerum*, 11:313.
32 Ibid., pp. 174-175.
33 Ibid., p. 204.
34 Ibid., p. 204, citing Pseudo-Paracelsus, *De natura rerum*, 11:317.

35   Ibid., p. 205.

36   Ibid., p. 217.

37   Ibid., pp. 217-218.

38   Ibid., p. 218.

39   Ibid., p. 222, emphasis added.

40   Ibid., p. 236.

41   Ibid.

42   No author, "Animal-Human Hybrids Spark Controversy," *National Geographic*, Thursday, October 28, 2010, http://news.nationalgeographic. com/news/2005/01/0125_050125_chimeras.html

43   Ibid., emphasis added.

44   Ibid.

45   Ibid.

46   Stephanie Feldstein, "Human-Animal Hybrids and Other Crimes Against Nature," http://news.change.org/stories/human-animal-hybrids-and-other-crimes-against-nature.

47   Rick Weiss, "Of Mice, Men and In-Between: Scientists Debate Blending of Human, Animal Forms," *The Washington Post*, November 20, 2004, http://www.infowars.com/articles/brave_new_world/chimera.htm.

48   Violet Jones, "Chimeras, Cloning, and Freak Human-Animal Hybrids," November 23, 2044, citing November 2001 CNN reports, at http:// www.infowars.com/articles/brave_new_world/chimera.htm.

49   Rick Weiss, "Of Mice, Men and In-Between: Scientists Debate Blending of Human, Animal Forms," *The Washington Post*, November 20, 2004, http://www.infowars.com/articles/brave_new_world/chimera.htm.

50   Ibid.

51   See Joseph P. Farrell, *The Cosmic War: Interplanetary Warfare, Modern Physics, and Ancient Texts* (Adventures Unlimited Press, 2007), pp. 139-150, *Genes, Giants, Monsters, and Men: The Surviving Elites of the Cosmic War and Their Hidden Agenda* (Feral House, 2011), pp. 125-161; *LBJ and the Conspiracy to Kill Kennedy: A Coalescence of Interests: A Study of the Deep Politics and Architecture of the Coup D'Etat to Overthrow Kennedy* (Adventures Unlimited Press, 2011), pp. 193-194, n. 6.

52   Rick Weiss, "Of Mice, Men and In-Between: Scientists Debate Blending of Human, Animal Forms," *The Washington Post*, November 20, 2004, http://www.infowars.com/articles/brave_new_world/chimera.htm.

53   Ibid.

54   Ibid.

55   Mary Shelley, *Frankenstein: The Original 1818 Text*, ed. D.L. Mcadonald & Kathleen Scherf, Second Edition (Broadview, 1999), pp. 82-83.

56   William R. Newman, *Promethean Ambitions*, p. 6, emphasis added.

57   James Shreeve, *The Genome War*, p. 227, cited in Joseph P. Farrell, *Genes, Giants, Monsters, and Men*, p. 136.

❧ Five ❧

FRANKENFOODS FOR THE "ALCHEMO-VEGETABLE" FRANKENSTEIN:

THE SEEDLESS SEEDS OF THE ANDROGYNE'S FOOD

❖

"Man is what he eats."
"If therefore my work is negative, irreligious, atheistic, let it be remembered that atheism at least in the sense of this work is the secret of religion itself..."
—Ludwig Feuerbach

"When (Percy Bysshe) Shelley reached Oxford, he poured out his thoughts concerning the possible uses of heat and combustion to transform matter, produce food, and eliminate starvation and slavery "
—James Bieri[1]

THE IDEA THAT IF ONE EATS a certain kind of sacred food that a transformation of mankind - from acquisition of special knowledge, to immortality, or, conversely, death - will result is as old as the Vedas and the Old Testament. It is darkly revealing that it is at the level of the second alchemical ascension - the "androgynous" fusion of man and vegetable into some sort of monstrous "alchemo-vegetable" creature - that one encounters the activity of powerful banking and corporate interests and of their lackeys who, in turn, have their own explicit and evident esoteric and occult interests. But this should not surprise us, for the doctrine of man as a microcosm, and of the universe as a "great man" (or μακανθρωπος), implies that for the total alchemical transformation of man, both his body, and his "extended" body

- i.e., the world, society, or environment - in which he lives, must likewise be transformed, including all that enters it, namely, his food. As we shall discover, however, it is not just mankind's food that is being tampered with, as the techniques of genetic engineering have already been applied to the creation of human-plant hybrids, raising the question, once again, of just exactly what constitutes a human being in law.

The alchemical roots of this modern genetic transformation may not, however, be immediately clear without a little background. The fusion of the vegetable with the human was, as we saw in chapter two, part of the core doctrine of the descent of man from a primordial androgynous unity. But this was, for many ancient societies, more than just a tenet of a metaphysical system. The Mayans, for example, attempted to practice a version of the fusion of plant and human by "fertilizing" their corn fields with the blood of their human sacrificial victims.[2] But this was not all there was to it.

Corn, along with wheat, is one of the seven sacred grains of alchemy, was indeed sacred to many ancient cultures, and is, of course, a staple grain in mankind's diet. Alchemically, it "symbolized the eternal return of life and the abundance of nature."[3] By so attempting to "fertilize" corn with human blood - including even the blood from the king's penis[4] - the Mayans were basically trying to effect the alchemical transformation of man by reactivating the fusion with the plant world.

A. The Alchemical Background:
the Rockefellers and Francis Bacon

The doctrine of man as a microcosm - a "little universe" - mirroring in himself the entities and relationships in the larger universe (viewed as a "makanthropos" or "large man") found expression in late mediaeval and Renaissance alchemical and esoteric medical practice in the notion that illnesses or diseases were not "punishments for sin," but were simply an imbalance or dissonance in the harmonies that were presumed to exist between man and the universe.

As man descended from androgyny, through the mineral, vegetable, and animal realms, a part of him was always immediately connected to each of those realms, and ultimately to the physical medium, the "androgynous aether," itself. Thus, each man possessed, according to this doctrine, a kind of "aetheric double" or spiritual component. Consequently, to cure a disharmony or disease meant to treat not just the physical symptoms but to restore the entirety of correspondences of mankind in all his levels.

This finds expression, for example, in the thinking of Paracelsus:

114.

Paracelsus, recognizing derangements of the etheric double as the most important cause of disease, sought to reharmonize its substances by bringing into contact with it other bodies whose vital energy could supply elements needed, or were strong enough to overcome the diseased conditions existing in the aura of the sufferer. Its invisible cause having been thus removed, the ailment speedily vanished.

The vehicle for the *archoeus*, or vital life force, Paracelsus called the *mumia*. A good example of a physical mumia is vaccine, which is the vehicle of a semi-astral virus. Anything which serves as a medium for the transmission of the archaeus, whether it be organic or inorganic, truly physical or partly spiritualized, was termed a mumia.[5]

In effect, what this implies is that in "alchemical medicine," one could seek to cure disease by the ingestion of harmony-restoring mineral, vegetable, and animal substances, or the manipulation of those substances in the wider environment, or both. And of course, it also meant that the "most universal form of the mumia" was ether itself, the physical medium, the Philosophers' Stone or Elixir, "which modern science has accepted as a hypothetical substance serving as a medium between the realm of vital energy and that or organic and inorganic substance."[6]

If all this sounds about as far removed from modern social engineering, or the genetic manipulation of man's food, it is not, for this basic alchemical view, we believe, is also capable of being reduced to a crassly material form, and the people that took this approach was a very famous - or, depending upon one's lights, infamous - bankster family, the Rockefellers.

Researcher Philip Regal described this "Rockefeller Corollary" of the alchemical pursuit of the transformation of mankind in no uncertain terms:

From the perspective of a theory reductionist, it was logical that social problems would reduce to simple biological problems that could be corrected through chemical manipulation of soils, brains, and genes. Thus the Rockefeller Foundation made a major commitment to using its connections and resources to promote a philosophy of eugenics.

The Rockefeller Foundation used its funds and considerable social, political, and economic connections to promote the idea that society should wait for scientific inventions to solve its problems, and that tampering with the economic and political systems would not be necessary. Patience, and more investment in reductionist research would bring trouble-free solutions to social and economic problems.[7]

But that was not all. As Regal also noted, "the project was in the general spirit" of that consummate fence-straddler between modern science and alchemy, Francis Bacon, and his *New Atlantis* "and Enlightenment visions of a trouble-free society based on mastery of nature's laws and scientific/technological progress."[8]

The reference to Bacon's *New Atlantis* deserves some comment. This short and highly esoteric work is usually considered to be part of a larger work, his *Advancement of Learning*, one of his many treatises on science. Bacon explicitly states in that work that there was a high science in antiquity, and that this in turn allowed the ancients to navigate the world much more easily than standard history allowed:

> You shall understand (that which perhaps you will scarce think credible) that about three thousand years ago, or somewhat more, the navigation of the world (especially for remote voyages) was greater than at this day. Do not think with yourselves, that I know not how much it is increased with you, within these six-score years; I know it well, and yet I say, greater then than now; whether it was, that the example of the Ark, taht saved the remnant of men from the universal deluge, gave men confidence to adventures upon the waters, or what it was; but such is the truth. The Phoenicians, and specially the Tyrians, had great fleets; so had the Carthaginians their colony, which is yet further west. Toward the east the shipping of Egypt, and of Palestine, was likewise great. China also, and the great Atlantis (that you call America), which have now but junks and canoes, abounded then in tall ships."[9]

Bacon, in other words, is subtly implying that America was to be the land of his grand social experiment to test the ways of magic and science in the production of a harmonious society. Worse yet, Bacon in his parable explicitly states that the inhabitants of the America's were "the descendants of Neptune planted there,"[10], and even states that "the several degrees of ascent, whereby men did climb up" to a magnificent temple, "as if it had been a Scala Caeli"(Ladder to Heaven) was located there.[11]

But what was the origin of all this navigational ability and the science that it represented? In Bacon's parable, the king of Atlantis established a kind of "scientific think tank," called "Salomon's House" which was "dedicated to the study of the works and creatures of God."[12] This "think tank," moreover, was found "in ancient records" as an "order" or "society" instituted "for the finding out of the true nature of all things."[13] In other words, Bacon is reproducing the esoteric tradition that after the Deluge and associated "cosmic catastrophes," an institution was founded to preserve and expand upon the knowledge of whatever civilization pre-existed the catastrophe.

In words that could almost be used to describe the goals and agenda's of the Rockefeller Foundation, Bacon describes the goals of "Salomon's House" in sweeping and universal terms: "The end of our foundation is the knowledge of causes, and secret motions of things; and the enlarging of the bounds of human empire, to the effecting of all things possible."[14] The goal, in other words, of science was to render the memes and myths of esoteric and alchemical lore a reality:

> For the several employments and offices of our fellows, we have twelve that sail into foreign countries under the names of other nations (for our own we conceal), who bring us the books and abstracts, and patterns of experiments of all other parts. These we call Merchants of Light.
>
> We have three that collect the experiments which are in all books. These we call Depredators.
>
> We have three that collect the experiments of all mechanical arts, and also of liberal sciences, and also of practises which are not brought into arts. These we call Mystery-men.
>
> We have three that try new experiments, such as themselves think good. These we call Pioneers or Miners.
>
> We have three that draw the experiments of the former four into titles and tables, to give the better light for the drawing of observations and axioms out of them. These we call Compilers.
>
> We have three that bend themselves, looking into the experiments of their fellows, and cast about how to draw out of them things of use and practice for man's life and knowledge, as well for works as for plain demonstration of causes, means of natural divinations, and the easy and clear discovery of the virtues and parts of bodies. These we call dowry-men or Benefactors.
>
> Then after divers meetings and consults of our whole number, to consider the former labours and collections, we have three that care out of them to direct new experiments, of a higher light, more penetrating into Nature than the former. These we call Lamps.
>
> We have three others that do execute the experiments so directed, and report them. These we call Inoculators.
>
> Lastly we have three that raise the former discoveries by experiments into greater observations, axioms, and aphorisms. These we call Interpreters of Nature."[15]

All of this, it is to be remembered, is being done *covertly* and by *Atlantis*, i.e., by America. America, in other words, was to become the great laboratory for a grand esoteric experiment being run by a hidden and ancient elite, which, let it be noted, Bacon is also describing in terms that are also applicable to the idea of a "breakaway civilization."

B. Esoteric Eugenics, Banksters, Seedless Seeds, and Alchemovegetable Man

But how does this universal Baconian scientism work itself out in terms of modern applications of the alchemical transformation of man? As we have already noted, between Paracelsus, Francis Bacon, and modern science, there is a progressive reduction of the alchemical agenda to purely materialist causes and techniques, and this is nowhere more in evidence than in the modern pursuit to create alchemo-vegetable man.

Four presuppositions guide this pursuit by contemporary elites:

1) That the world has an upper limit "carrying capacity" for population, i.e., that it is overpopulated with too many "useless eaters";
2) That this calls for social engineering measures, including various "methods" of population reduction and the genetic manipulation of the food supply;
3) That this in turn can be effected by the engineering of man himself by the injection of human genes into plants; or,
4) by the injection of plant genes into humans.

As we shall outline in the subsequent sections of this chapter, points 2-4 above also carry with them legal consequences for humanity.

1. A Babylonian Theme Revisited: Too Many People (?)

The idea that the earth is overpopulated, and that the elite must "do something about it" was first expressed in ancient Mesopotamian text, the *Atrahasis Epic*. In that epic, the god Enki and the goddess Mami create humanity, at the behest of the other gods, as a slave-worker, from a pre-existent hominid, a female, and one of the "gods," a male. Thus, in the Babylonian view, modern mankind is already a hybrid or chimerical creature, part "divine" and part "human." There were, however, two problems. The new chimerical man bred too quickly, and lived far too long, thus overpopulating the earth and by sheer numbers, threatened the power of the "gods." The god Ellil, great rival to Enki, then complained that mankind was making too much noise, and he and the other gods determine to wipe out the vast bulk of their own creation. In other words, mankind was guilty of "noise pollution". And the methods the gods choose to wipe out mankind are, for our purposes, intriguing, for they first try diseases, then starvation, before finally deciding to send a flood.[16]

2. The Vipers of Venice Reiterate the Theme:
Carrying Capacity

This very Babylonian idea migrated to Italy via the Roman Empire's conquest of the region, and its consequent importation of "Chaldean" slaves to the suburbican dioceses of Italy itself. Following the Roman custom, many of these slaves were freed upon the deaths of their owners, and they became important members of the late Roman imperial bureaucracy, and when the Western Empire finally collapsed, eventually became the roots of the powerful oligarchical banking families of the "Serene Republic," Venice.[17]

There, according to researcher Webster Tarpley in a landmark study, *Against Oligarchy*, the Venetan financiers, facing the collapse of their "Serene Republic," backed a study by a scholar, Gianmaria Ortes, who declared that the earth had a maximum carrying capacity of three billion people.[18] Of course, Ortes' figure as long since been exceeded, but it is worth noting that, upon the collapse of the Venetian Republic, its financial elite simply "transferred shop" to another "city in a swamp," Amsterdam, and from thence ultimately to London, carrying their overpopulation ideas with them to the modern era, and such exponents of overpopulation theories as Thomas Malthus, and, of course, his modern American counterparts in the financial elite. And here, as they say, is where the story gets interesting, for once again, in an almost verbatim replay of the ancient Babylonian *Atrahasis Epic*, mankind is viewed as a form of "overpopulation pollution."

3. The Banksters Adopt the Babylonian Theme

It is now a known and established fact that the Rockefeller Foundation financed the American Eugenics Society and Record Office in Cold Harbor New York.[19] There, "millions of index cards on the bloodlines of ordinary Americans were gathered: with a view to "map the inferior bloodlines and subject them to lifelong segregatioon and sterilziation to 'kill their bloodlines.'"[20] This should not surprise us, since, as research F. William Engdahl observed, John D. Rockefeller III was raised in an environment where he was constantly surrounded by "eugenicists, race theorists, and Malthusians" such as Alan Greg,[21] who viewed the slums of America's inner city as a form of cancerous nekrosis, as a kind of pollution or disease to be eradicated[22] or, as Engdahl summarizes these views, a herd to be culled.[23]

One of these Rockefeller "experts" was Frederick Osborne, who became the first President of John D. Rockefeller III's Population Council, and who had been "a founding member of the American Eugenics Society."[24] As if now well-known, the views of these early eugenicists were codified into law in Nazi Germany's Nuremberg Race Laws of the 1930s. But after the horrors of World War Two had fully revealed the extent of Nazi atrocities and genocide, this left the Rockefeller Malthusians with something of a problem, for their program had to be re-packaged, for by backing the Nazis, the same eugenicists, as Osorne wrote in 1956 for *Eugenics Review*, "all but killed the eugenic movement."[25] Osborne whined "People are simply not willing to accvept the idea that the genetic base on which their character was formed is inferior and should not be repeated in the next generation.... They won't accept the idea that they are in general second rate...."[26] Consequently, the whole agenda had to be re-packed and resold:

> Eugenics was to be mass-marketed under a new guise. Instead of talking about eliminating "inferior" people through forced sterilization or birth control, the word would be "free choice" of family size and quality.[27]

All of this was to be accomplished by a mass-market campaign of social engineering, appealing to a new idea, that of "wanted children".[28]

As if all this were not enough, Engdahl notes that in May of 1932, "the Rockefeller Foundation sent a telegram to its Paris office, which quietly funnelled the US Rockefeller funds into Germany."[29] There the funds were to be used to study twins for the purposes of studying "effects on later generations of substance toxic for germ plasm," in other words, a genetic-specific bio-weapon,[30] and all this a year before the Nazis took power, who would make such studies a part of its grizzly inventory of "medical research" in the concentration camps. And what of Osrborne? After Watson and Crick cracked the structure of the DNA double helix, making "scientific eugenics" truly feasible, Osborne qouted a Nazi eugenics researcher, Hermann J. Müller, on what the real goal was: "It would in the end be far easier and more sensible to manufacture a complete new man de novo out of appropriately chosen raw materials, than to try to refashion into human form those pitiful relics which remained."[31] To this end, Osborne also approved of Müller's idea to found "sperm banks" in order to "make available the sperm of highly qualified donors."[32]

4. The Rockefellers, the "Food Weapon", and the Alchemical Seedless Seeds
a. A World War, and the "Peace Studies Group"

But where does food fit in?

The answer may be summed up in one word, or rather, one surname: Rockefeller.

This aspect of the story begins, as so many do, in the early days of World War Two, when the Rockefeller Foundation funded a private and secretive policy group, the War and Peace Studies Group of the New York Council on Foreign Relations, to the tune of $350,000.[33] The purpose of the study group was to determine the post-war role of the USA and "to shape US post-war economic and political goals, based on the assumption a world war would come and that the United States would emerge from the ashes of that war as the dominant global power."[34] The vision this study group projected was that the USA would replace the British Empire as the pre-eminent global power, and it planned accordingly. Rather than basing its vision of the postwar world on actual physical possession of colonies as in the British model, the CFR study group based its vision around the exercise of controlling economic influence. "It was," notes F. William Engdahl, "a brilliant refinement which allowed the US corporate giants to veil their interests behind the flag of democracy and human rights for 'oppressed colonial peoples,' support of 'free enterprise' and 'open markets.'"[35]

Within this context, John D. Rockefeller III was, of course, pursuing his eugenics and depopulation policies via his Population Council, while his brother Nelson "was working the other side of the fence,"[36] seeking new methods of increasing the efficiency of worldwide food production .[37] It is within this context that agriculture and food became an instrumentality of policy, and, a weapon. In short, it was a postwar goal of this group to have the US, both government and corporations, dominate agricultural technology and thereby global food production.

To this end, the Rockefeller interests began to promote a "Green Revolution," promoting increased agricultural efficiency verses "communist inefficiency" across a variety of nations that were seen as "sensitive": India and other nations in Asia, Mexico and other Latin American nations, as we shall see shortly. This increase of production efficiency had the result of driving many peasants from their land as they were no longer needed in the new emerging world of "agribusiness," driving them into the inner city slums, where they would be cheap labor for the giant US multinational companies. This, of course, was part of a deliberate plan.[38]

b. An Esoteric Connection?

This deliberate targeting of the agriculture of developing countries began in 1941, a few months before Pearl Harbor, when Nelson Rockefeller and US Vice President Henry Wallace - himself a leading figure in emerging "agribusiness" with his strong stock interests in the company that would later become an agribusiness giant, Pioneer Hi-Bred company, which would later become a component of the large DuPont agribusiness empire[39] - sent a team to Mexico to conduct discussions with the Mexican government on increasing food production.

It is worth mentioning that this was the same Henry Wallace who was a high-ranking Freemason, and who convinced fellow Mason, President Franklin Roosevelt, to place the esoteric symbolism of the Great Seal of the United States, with its uncapped pyramid and Eye of Horus, and other occult symbols, on the obverse of the one dollar bill.[40]

Yet another prominent public figure with strong ties both to esotericism and to the Rockefeller Empire is Maurice Strong, "the Rockefeller family's international environmental organizer" and Rockefeller Foundation Trustee.[41] Strong and his wife helped to found the "spiritual center," essentially a kind of ecumenical ashram of Hindu, Buddhist, and even a Catholic Carmelite center, near the small town of Crestone, Colorado.

What these esoteric and religious interests might indicate is that the esoteric agenda - particularly in Wallace's case and is connection to the Rockefeller interests - may be playing a hidden role within the development of the postwar goals of American agribusiness.

As we shall now see, the technologies being developed raise that possibility considerably.

c. The" Food Weapon" and Other Techniques of Alchemical Social Engineering

We have repeatedly observed in the previous pages that one goal of alchemy has always been the transformation of human consciousness, or, in modern terms, to "manage perceptions," and nowhere is this agenda more in evidence than in the strategy employed by the Rockefeller interest to manage the perceptions of scientists engaged in the genetic studies so essential to agribusiness:

John D. Rockefeller III's Agricultural Development Council also deployed US university professors to select Asian universities to train a new generation of scientists. The best scientists would then be selected to be sent to the United States to get their doctorate in agriculture sciences, and coming out of the American universities, would follow the precepts close to the Rockefeller outlook on agriculture. This carefully-constructed network was later to prove crucial in the Rockefeller Foundation's subsequent strategy to spread the use of genetically-engineered crops around the world.[42]

In short, the Rockefeller Foundation because the keystone in a vast arch of social engineering, via its grants to train hundreds of scientists around the world, thus managing their perceptions, socially engineering a "scientific culture" in which an "agri-world view" favorably disposed to genetically modified crops would prevail.[43]

Genetic modification of the food supply was a key goal of the Rockefeller interests, for having realized that "science would eventually come to control the fundamental processes of biology," those associated with "the Rockefeller institutions saw it as the ultimate means of social control and social engineering, eugenics."[44] To this end, the Rockefeller Foundation invested hundreds of millions of dollars in the 1980s, both directly and indirectly, to sponsor genetic research into food crops.[45] As a component of this research, it had specifically targeted the "rice bowl" of Asia, and the genetic mapping of the rice genome.[46] The result of this was that centuries of agronomical experience in creating natural diversities of rice species in a period of a few decades, and "drew Asia's peasantry into the vortex of the world trade system and the global market for fertilizer, high-yielding seeds, pesticides, mechanisation, irrigation, credit and marketing schemes packaged for them by Western agribusiness."[47] The effect of these sorts of projects was to transform farmers the world over, from America to Asia, into something very similar to a feudal serf, "indentured through huge debts, not to a Lord of the manor, but to a global multinational corporation such as Cargill, Archer Daniels Midland, Smithfield Foods or ConAgra."[48] The reason for this transformation was very simple: patent law. As agribusinesses expanded their inventories and sales of genetically modified crops, farmers quickly discovered that centuries of agronomy were thrown by the wayside. Seeds could no longer be retained from one harvest to seed the next, for the newer crops were patented, and the licensing agreements from the seed companies to the farmers prohibited this centuries' old practice.

A prime example of this strategy was Iraq. Here we must cite Engdahl at length, for his remarks cast a very long shadow of recent and current events in the Middle East:

Iraq is historically part of Mesopotamia, the cradle of civilization, where for millennia the fertile valley between the Tigris and Euphrates rivers created ideal conditions for crop cultivation. Iraqi farmers have been in existence since approximately 8,000 B.C., and developed the rich seeds of almost every variety of wheat used in the world today. They did this through a system of saving a share of seeds and replanting them, developing new naturally resistant hybrid varieties through the new plantings.

For years, the Iralis held samples of such precious natural seed varieties in a national seed bank, located in Abu Ghraib, the city better known internationally as the site of a US military torture prison. Following the US occupation of Iraq and its various bombing campaigns, the historic and valuable seed bank in Abu Ghraib vanished, a further casualty of thr Iraq war.

However, Iraq's previous Agriculture Ministry had taken the precaution to create a back-up seed storage bank in neighboring Syria, where the most important wheat seeds are still stored in an organization known as the International Center for Agricultural Research in Dry Areas (ICARDA), based in Aleppo, Syria.[49]

These seeds could have been used as the basis for a new Iraqi food crop, but the corporate-backed US occupation authority prohibited their use. Iraqi farmers were forced to buy the genetically modified crops of American corporations, rather than seeds derived from millennia of agronomical practice.[50] We will return to this important point of the distinction between agronomy and genetically engineered crops below.

Iraq had become the latest in a long line of agricultural alchemical laboratories for social engineering via the introduction of genetically modified crops. But it was not the first...

d. Argentina:
The Alchemical Laboratory for the
Seedless Seeds of Alchemy
(1) The Historical Background: the Beginning:
Rockefellers, Nazis, and Perón

Argentina was a central stage in the postwar plans of the American corporate elite, and, as Joseph has documented elsewhere, in the postwar plans of another elite: the Nazis, and both planned to work in concert there with each other. The Rockefeller interests were intimately involved in assisting their

Nazi business partners in I.G. Farben to move Nazi loot into various banks, but this story is too big to be told here.[51]

The Rockefeller interest in Argentina began during World War Two, when Nelson Rockefeller was the coordinator of US intelligence and covert operations in Latin America. In this role, Nelson was able to persuade Argentine President Juan Domingo Perón to declare war on Nazi Germany in March 0f 1945. But this should not be taken as implying that Perón was breaking with his friendship for the Nazis, for once again, it was part of a deliberate policy. As the Rockefeller banks were helping the Nazis launder their money out of Europe, Perón, as he himself noted in his memoirs, was able by dint of his declaration of war under international law to enter Germany, and assist his Nazi "enemies" to escape to Argentina:

> The false declaration of war...

a false declaration, let us remember, that Nelson Rockefeller persuaded Perón to make,

> ...had a clear purpose: "We hadn't lost contact with Germany, despite the break in diplomatic relations," Perón would say in 1967. "Things being so we received an unusual request. Even though it may seem contradictory at first, Germany benefits from our declaration of war: if Argentina becomes a belligerent country, it has the right to enter Germany when the end arrives: this means that our planes and ships would be in a position to render a great service. At that point we had the commercial planes of the FAMA line (Argentine Merchant Air Fleet) and the ships we had bought from Italy during the war. That is how a great number of people were able to come to Argentina.[52]

Under this doubly malign influence of American bankers and Nazis, Argentina was eventually to be transformed into a huge laboratory for a Rockefeller "social experiment," in addition to being a haven for postwar Nazi secret research.

(2) The Rockefellers and Argentina

The Rockefeller side of this story began in the early 1980s. Under Perónism, Argentina's agriculture and family farmers had prospered, producing high quality beef while raising small amounts of crops on family farms passed down through the generations. Argentina's agriculture was self-sustaining, and was a major agricultural exporter.[53]

Then came the Yom Kippur War of 1973, and its aftermath, the artificially engineered oil crisis of the 1970s. This allowed the large international banks to sell loans to nations such as Argentina in order to assist them to import oil. The terms of the loans were generous enough to tempt those nations into accepting the loans, which Argentina did.

Then, the banksters struck again. In 1979, faced with a collapsing dollar, the US Federal Reserve raised its interest rate "by some 300%, impacting worldwide interest rates, and above all the floating rate of interest on Argentina's foreign debt."[54] The consequence for Argentina was such that it was "caught in a debt trap not unlike that which the British had used in the 1880's to take control of the Suez Canal from Egypt. New York bankers, led by David Rockefeller had learned the lessons of British debt imperialism."[55] They had not only learned those lessons, but had learned them well.

All of this was aided by bringing to an end the Perónist structure of Argentina, with the Washington-Rockefeller-backed military coup in 1976. The new regime proved to be a problem, however, for it was

> too liberal in its definition of human rights and due process of law. In October 1976, Argentine Foreign Minister, Admiral Cesar Gizzetti met with Secretary of State Henry Kissinger and Vive President Nelson Rockefeller in Washington. The meeting was to discuss the military junta's proposal for massive repression of opposition in the country. According to declassified US State Department documents released only years later, Kissinger and Rockefeller not only indicated their approval, but Rockefeller even suggested specific key individuals in Argentina to be targeted for elimination.[56]

But why, beyond Argentina's relative prosperity under its Perónist system, was it the focus of such attacks by the American corporate elite?

The answer is that the agribusiness elite intended for it to become "a secret experimental laboratory for developing genetically engineered crops. The population was to become the human guinea pigs of the project."[57] Under the regime of President Menem, who replaced the junta (again with the maneuverings of the American government and agribusiness elite in the background), the Argentine government granted more than 569 licenses for field trials for various genetically engineered crops, including "corn, sunflowers, cotton, wheat and especially soybeans."[58]

That brings us at last to the final alchemical transformations, not only of food, but of society.

C. Genetically Modified Crops and The Patent Weapon:
Patented Plants, Pigs, and, Maybe People?
1. Secret Meetings and "Substantial Equivalence"

While the Rockefeller interests were promoting "agri-business" and genetically modified foods through their private financial empire, a coordinated effort was also underway within the Reagan Administration, the key figure of which was then Vice President, and later President, George Herbert Walker Bush, the former National Republican Party Chairman and CIA director. In 1986, Vice President Bush hosted a "special White House strategy meeting" with executives from the American corporate giant, Monsanto, to map out plans for deregulation of the biotech industry.[59]

Adopting the strategy that traditional methods of animal breeding and plant breeding - agronomy - were "substantially equivalent" to the newer techniques of genetic engineering,[60] This Committee laid down the policies that allowed the agribusiness industry to by-pass emerging concerns from scientists - many of them within government, academia, and the corporate world - that the genetically modified foods required stricter regimens of testing.[61]

But the real contradiction, and the almost magical way in which these concerns were by-passed, lay in the principle of "substantial equivalence" itself. Genetic modification of plants is a far cry from the agronomic method of selective breeding of certain qualities over several generations, for genetic modification often involves a kind of "shotgun" approach, bombarding a plant with selected bacteria, or alternatively, selecting certain characteristics within a genome of one plant (or animal) and splicing them into the target plant.

However, while large agribusiness companies were busily genetically modifying crops and arguing against strict testing regimens for their creations on the one hand, they revealed the essential *non*-equivalence of the two methods - traditional agronomy versus direct genetic modification - by claiming the right to exclusive patent protection for their concoctions.[62] Indeed, they could claim a certain dubious basis in American patent law, since these creations could not otherwise have arisen in nature, and were accomplished only by the intervening hand of man.[63]

But why were these corporations so intent upon securing their patent rights over their new hybrid seeds? The answer, and the dark shadow it casts, is simple, and Engdahl's explanation cannot be improved upon:

> Hybrids had a built in protect against multiplication. Unlike normal
> open pollinated species whose seed gave yields similar to its parents,

the yield of the seed borne by hybrid plants was significantly lower than that of the first generation.

That declining yield characteristic of hybrids meant farmers must normally buy seed every year in order to obtain high yields. Moreover, the lower yield of the second generation eliminated the trade in seed that was often done by seed producers without the breeder's authorization. It prevented the redistribution of the commercial crop seed by middlemen. If the large multinational seed companies were able to control the parental seed lines in house, no competitor or farmer would be able to produce the hybrid.[64]

As a result, these companies, by realizing the old alchemical goal of "seedless seeds," saw the Rockefeller goal of dominating the world's food supplies within their grasp.[65] Indeed, Monsanto took out a U.S. patent, number 5,723,765, "Control of Plant Gene Expression," which was a patent on the very *concept* of modifying *any* plant "to kill its own embryos" and was applicable to all plants and seeds of any species.[66] As French researcher Marie-Monique Robin put it, substantial equivalence "represents the nub" of "one of the greatest conspiracies in the history of the food industry."[67]

Robin also cites the Consumers Union advocate, Michael Hansen, who points out the double standard, double talk, and inherent contradiction of substantial equivalence and the behavior of the corporations advocating it:

> We have always criticized the doubletalk of biotechnology companies... On the one hand, they say there is no need to test transgenic plants because they are exactly the same as their conventional counterparts; on the other, they file for patents, on the grounds that (genetically modified organisms) are unique creations. You have to make up your mind: either (genetically modified) soybeans are identical to conventional soybeans, or else they're not. They can't be both depending on Monsanto's interests.[68]

Such measures allowed the multinational agribusiness corporations to charge license and royalty fees on the use of their seeds,[69] and even to sue farmers who, not having planted their genetically modified seeds, were nevertheless found to have fields growing these hybrids, brought there by entirely natural means! Such measures also allowed these seeds, once introduced into an area, oftentimes with tacit or even official US government support, to coerce other nations into accepting policies dictated in Washington, D.C. It was a classic "food or famine" strategy, and it was effective.[70]

In other words, genetic engineering was the means to the end, the end being to gain patent rights over plants being used in agricultural production; it was a means, in short, of the social transformation of mankind.[71] Patents, in other words, became the instruments of conquest, and this, Robin points out, was their actual origin in jurisprudence:

> The word 'patent' itself comes from the age of conquest. 'Letters patent' was the name given to an official public document - in Latin, *patens* means 'open' or 'obvious' - bearing the seal of European sovereigns [and] granting to adventurers and pirates the exclusive right to conquer foreign countries in their name. At the time Europe was colonizing the world, letters patent were directed at territorial conquest, whereas today's patents are aimed at economic conquest through the appropriation of living organisms by the new sovereigns, the multinational corporations like Monsanto.[72]

But surely American patent law cannot be made to extend to the rest of the world, right?

Wrong.

While the agribusiness multinationals were transforming Canada, Argentina, India and the USA into laboratories for their products, they were also extending their right to enforce patents, and thereby, royalties fees, on the farmers planting them. This was through the TRIPS, or Trade Related Aspects of Intellectual Property Rights, agreement, sponsored by yet another extension of the American government-corporate nexus, the WTO, or World Trade Organization. Traveling to Geneva in 2005, Marie-Monique Robin asked a question of Adrian Otten, then the director of intellectual property for the World Trade Organization. At the beginning of her interview, Robin

> asked a question that suddenly made him tense up: "What is the goal of the TRIPS agreement?" Stammering a bit, he finally answered, "Well, I suppose that one of the fundamental objectives is to establish common international rules of member governments of the WTO to protect the intellectual property rights of certain member countries of the WTO, as well as those of their citizens and companies."
>
> "And which article has caused a problem? I asked, to see if I had understood the WTO's gibberish.
>
> "Well, it's Article 27, paragraph 3(b), which adds a clause to the TRIPS agreement according to which inventions connected to plants and animals should be subject to patenting."

Put like that, it was a clear as spring water.

"The goal of the TRIPS agreement is that a patent obtained in the United States - for example, by Monsanto - will be automatically applicable everywhere in the world," I had been told a month earlier in New Delhi by Devin-der Sharma. Chairman of the Forum for Biotechnology and Food Security, this noted Indian journalist is a fierce opponent of the WTO. "If you observe the international evolution of the patent system, you can see that it follows exactly the Patent Office in Washington. With the TRIPS agreement, every country has to follow the model of the United States or else suffer severe commercial penalties, because the WTO has absolutely extraordinary powers of coercion and reprisal.... The TRIPS agreement was also designed by multinational corporations to seize the genetic resources of the planet, chiefly in Third World countries, which have the greatest biodiversity. India is a particular target, because it is a megadiverse country where there are 45,000 plant species and 81,000 animal species.[73]

Not only did genetic engineering equal corporate engineering transforming the very nature of science itself from an "objective" pursuit of scientific fact, into a culture where certain views were promoted, while contradictory views were sidelined,[74] the paradigm was to be extended, via TRIPS and the WTO, to a global extent.

There is a possibility - albeit a slight one - raised by all this genetic engineering, this alchemical engineering of seedless seeds and its accompanying alchemical social engineering, and that is the possibility that, through some mutation, the genetic modifications of engineering foods might actually modify the genetic makeup of humans eating it.

And that in turn, raises the prospect - if the engineered genes are patented, that humans so modified might be considered the intellectual property of the corporations having produced the modification.

But, as we shall now see in the next chapter, there are other possibilities that the transhumanists are exploring, that might beat mother nature to the punch.

ENDNOTES

1   James Bieri, *Percy Bysshe Shelley: A Biography* (John Hopkins University Press, 2008), p. 51.
2   "Corn," in *Alchemical Properties of Foods*, www.alchemylab.com/ guideto.htm.
3   Ibid.
4   Ibid.
5   Manley P. Hall, *The Secret Teachings of All Ages, Reader's Edition*, p. 347.
6   Ibid.
7   Philip Regal, *Metaphysics in Genetic Engineering: 2.2 Utopianism*, Beunoes Aires, 1996, http:www.psrast.org/pjrbiosafety.htm. Cited in F. William Engdahl, *Seeds of Destruction: The Hidden Agenda of Genetic Manipulation* (Global Research, 2007), p. 157.
8   Ibid.
9   Francis Bacon, Lord Verulam, *The Advancement of Learning and The New Atlantis* (Oxford University Press, 1966), p. 271.
10 Ibid., p. 272.
11 Ibid.
12 Ibid., p. 276.
13 Ibid., p. 277.
14 Ibid., p. 288.
15 Ibid., pp. 296-297.
16 For a fuller discussion of this ancient epic, see Joseph's *The Cosmic War: Interplanetary Warfare, Modern Physics, and Ancient Texts* (Adventures Unlimited Press, 2007), pp. 144-147.
17 For a fuller exposition of this story, see Joseph's *Babylon's Banksters: The Alchemy of Deep Physics, High Finance, and Ancient Religion* (Feral House, 2010),pp. 270-272.
18 Webster Tarpley, "Gianmaria Ortes: The Decadent Venetian Kook Who Originated the Myth of 'Carrying Capacity,'" *Against Oligarchy*, http://tarpley.net/online-books/against-oligarchy/giammaria-ortes-the-decadent-venetian-kook-who-originated-the-myth-of-carrying-capacity/. The role of Venice in the story of the structure and power of financial elites will be explored by Joseph in a future work.
19 F. William Engdahl, *Seeds of Destruction: The Hidden Agenda of Genetic Manipulation* (Global Research, 2007), p. 75.
20 Ibid., p. 77.
21 Ibid., p. 72.
22 Ibid., p. 73.
23 Ibid., p. 72.
24 Ibid.,p. 88.
25 Ibid.
26 Cited in John Cavanaugh-O'Keefe, *The Roots of Racism and Abortion: An Exploration of Eugenics, Chapter 10: Eugenics after World War II,* 2000, http://www.eugenics-watch.com/roots/index.html. The problem was much worse than that. As Engdahl points out, one early German eugenicist, Dr Franz J. Kallmann, was involved in research... until it was discovered that he too was a "second rate" person, being partly Jewish. Kallman thus had to leave Nazi Germany in 1936 to pursue his eugenics projects in the USA, where his "American Society of Human Genetics later became a sponsor of the Human Genome Project."(Engdalh, op. cit., p. 94). Woops.
        Another intriguing figure was the well-known biologist, Dr. Detlev W. Bronk, who attended the founding meeting of the Rockefeller Population Council, and who was also president "of both the Rockefeller Institute and the National Academy of Sciences," and who was "sympathetic to the agenda of population control."(Engdahl, op. cit., p. 85) Some readers will also recognize Detlev Bronk as being one of the

first alleged members of the so-called Majic-12 UFO study group, according to the Majic-12 Cooper-Cantwheel documents, whose authenticity is hotly contested within the ufology community.

27 Engdahl, op. cit., pp. 88-89.

28 Ibid., p. 89. Engdalh also notes that John D. Rockefeller III "made Puerto Rico into a huge laboratory to test his ideas on mass population control beginning in the 1950s. By 1965, an estimated 35% of Puerto Rico's women of child-nearing age had been permanently sterilized... The Rockefeller's Population Concil, and the US Government Department of Health Education and Welfare - where brother Nelson was Under-Secretary - packaged the sterilization campaign.... Poor Puerto Rican women were encouraged to give birth in sanitary US-built hospitals where doctors were under orders to sterilize mothers who had given birth to two children by tying their tubes, usually without the mothers' consent."(Engdahl, op. cit.,p. 70). So much for a woman's right to choose.

29 Ibid., p. 81.

30 Ibid.

31 Ibid., p. 91, citing Frederick Osborne, *The Future of Human Heredity: An Introduction to Eugenics in Modern Society*, Webright and Talley, New York, 1968, pp. 93-104.

32 Ibid.

33 Engdahl, op cit.,pp. 102-103.

34 Ibid., p. 102.

35 Ibid., p. 103. Engdahl also notes that the "American domination of the world" after the end of World War Two "would be accomplished via a new organization, the United Nations, including the new Bretton Woods institutions of the International Monetary Fund and World Bank, as wel as the General Agreement on Tariffs and Trade(GATT)."(p. 106)

36 Ibid., p. 107.

37 Ibid.

38 Ibid., p. 128. Engdahl also observes that "The Green Revolution was typically accompanied by large irrigation projects which often included World Bank loans to construct huge new dams, and flood previously settled areas and fertile farmland in the process."(p. 129)

39 Ibid., p. 111.

40 David Ovason, *The Secret Symbols of the Dollar Bill: A Closer Look at the Hidden Magic and Meaning of the Money You Use Every Day* (Perennial Currents, 2004), pp. 15-16.

41 Engdahl, op. cit., p. 127.

42 Ibid., p. 128.

43 Ibid., p. 161. It is worth noting that "During the late 1930's, as the (Rockefeller) foundation was still deeply involved in funding eugenics in the Third Reich, it began to recruit chemists and physicists to foster the invention of a new science discipline, which it named molecular biology to differentiate it from classical biology."(p. 153) This it did to deflect mounting criticism against its funding of eugenics.

44 Ibid., p. 154.

45 Ibid., p. 153.

46 Ibid., p. 160.

47 Ibid., p. 162.

48 Ibid., p. 137.

49 Ibid.,p. 202.

50 Ibid.

51 See Joseph P. Farrell, *The Nazi International: The Nazis' Postwar Plan to Control Finance, Conflict, Physics, and Space* (Adventures Unlimited Press, 2008), pp. 247-350, and *Saucers, Swastikas, and Psyops: A History of a Breakaway Civilization: Hidden Aerospace Technologies and Psychological Operations* (Adventures Unlimited Press, 2011), pp. 141-178. See also Engdahl, op. cit., p. 109.

52 Farrell, *Nazi International*, p. 172, citing Uki Goñi, *The Real ODESSA: How Perón Brought the Nazi War Criminals to Argentina* (London: Granta Books, 2002), p. 24.

53 Engdahl, op. cit., pp. 176-177.

54 Ibid., p. 177.

55 Ibid.

56 Ibid., p. 178. Engdahl cited U.S. Embassy *Document #1976 Buenos 06130,* 20 September 1976, in Cynthia J. Arnson, ed., *argentina-United States Bilateral Relations,* (Washington, D.C.: Woodrow Wilson Center for Scholars, 2003), pp. 39-40. (See Engdalh, p. 194, n. 3)

57 Ibid., p. 182.

58 Ibid.

59 Ibid.,p. 4.

60 Ibid., p. 5. The principle of "substantial equivalence" has also been commented on extensively by French researcher Marie-Monique Robin, *The World According to Monsanto: Pollution, Corruption, and the Control of Our Food Supply: An Investigation into the World's Most Controversial Company* (The New Press, 2010), pp. 136-145. Robin also cited Sussex University professor, Erik Millstone, that the principle of "substantial equivalence" was itself ill-defined: "The concept of substantial equivalence has never been properly defined: the degree of difference between a natural food and its GM alternative before its 'substance' ceases to be acceptable 'equivalent' is not defined anywhere, nor has an exact definition been agreed by legislators. It is exactly this vagueness which makes the concept useful to industry but unacceptable to the consumer. Moreover, the reliance by policy makers on the concept of substantial equivalence acts as a barrier to further research into the possible risks of eating GM foods."(pp. 170-171) In other words, by the sheer alchemy of words, and nothing more, agronomical techniques were rendering equivalent to genetic engineering techniques.

61 Ibid., p. 6.

62 Ibid., p. 8.

63 For a further discussion of this issue and its ramifications in terms of the content of ancient texts and their apparent claims to the genetically engineered origins of modern man, see Joseph P. Farrell, *Genes, Giants, Monsters, and Men* (Feral House, 2011), pp. 136-137, 156-158. Engdahl notes that even a genetically modified pig has been patented. (p. 203)

64 Engdahl, op. cit. p. 130.

65 Ibid., p. 257.

66 Ibid.,p. 258.

67 Robin, *The World According to Monsanto*, p. 146. Robin, citing Michael Hansen, a consumer advocate, also observes that the principle of substantial equivalence we noting more than an alibi "created out of thin air to prevent (genetically modified organisms) from being considered at least as food additives, and this enably biotechnology companies to avoid the toxicological tests" normally required under law. (p. 147)

68 Ibid., pp. 201-201.

69 Engdahl, *Seeds of Destruction*, p. 188.

70 Ibid., pp. 267-268. It is worth noting that this strategy may backfire, and mightily so, for farmers in Canada have begun to bring lawsuits against Monsanto for introducing crops into their fields without their knowledge or consent.

71 Robin, *The World According to Monsanto*, pp. 310-311.

72 Ibid., p. 312.

73 Ibid., pp. 316-317.

74 See the discussion beginning on p. 135 in Robin's book.

⚛ Six ⚛

THE "ALCHEMO-MINERAL" MAN:
THE TRANSHUMANIST "TECHNO-ANDROGYNY"
AND THE APOCALYPSE

∵

"Any sufficiently advanced technology is indistinguishable from magic."
—Arthur C. Clarke[1]

"There is a secret known
To thee and to none else of living things
Which may transfer the sceptre of Heaven,
The fear of which perplexes the Supreme..."
Mercury to Prometheus, *Prometheus Unbound,*
—Percy Bysshe Shelley[2]

IF, AS ARTHUR C. CLARKE famously observed, "any sufficiently advanced technology is indistinguishable from magic," then it is equally true that the *goals* of modern science and technology are indistinguishable from alchemy and hermeticism. Those goals, for science and alchemy alike, are the mastery over the physical medium, the environment of man, and over mankind himself, and this is nowhere better illustrated than in the "transhumanist" speculations concerning what is being called "The Singularity."

If that term - "Singularity" - sounds apocalyptic, that's because it is, for the transhumanists intend, consciously and deliberately, to transform man, and to do so specifically by creating the "techno-androgynous fusion" of man and mineral, of man and machine, and to do *that* by the application of every technical artifice available or conceivable. They speak openly of an apocalyptic future transcending evolution human nature itself, and of a future, much of it already dawning, in which technologies allow mankind to engineer human evolution itself. It is, in short, a Frankenstein scenario, for as Joel Garreau, a researcher of the transhumanist agenda, put it, "The inflection point at which we have arrived is one in which we are increasingly seizing the keys to all creation."[3]

But what exactly, in the transhumanists' view, *are* these "keys to all creation"? And what is their history? Who, really, is ultimately behind their development? We already encountered, in the previous chapter, a partial answer to the last question, with the private corporate interest - represented by the Rockefeller Foundation - pursuing a wholesale scientific agenda to engineer a new mankind. But there is also another significant player in the development of these technologies. But again, what *are* these technologies? And who else is behind their development?

A. DARPA's GRINs: A Brief Review of the Background of Transhumanist Technologies
1. The Keys to Creation

Transhumanists call the technologies that are "the keys to creation" the "GRIN" technologies, standing for genetic, robotic, information processing, and nano-technological processes.[4] These, the transhumanists stress, overlap and may be combined in any number of ways, and it is precisely this inter-mingling of various technologies that is "creating a curve of change unlike anything we humans have ever seen."[5] It is important to stress an obvious, though often over-looked point, about these four technologies, namely, that they are each, taken individually, transformative and each hold the potential to change human nature itself. Taken together, or engineered and employed in various combinations,[6] the transformation is even more sweeping.

Succinctly put, the GRIN technologies are transformational; they are *alchemical.*

Indeed, Joel Garreau points out that each of these technologies is aimed not just at the "surface" of mankind, but - to employ that topological meta-phor once again, at his *interior*, at his very essence or nature, such that it is possible to envision the technological fusions of "alchemo-mineral man", the fusion of man and mineral, of man and *machine,* the next highest fusion in our ladder of alchemical ascent, for these technologies allow mergers with "our minds, our memories, our metabolisms, our personalities, our progeny and perhaps our souls"[7] to the extent we are now capable of engineering our own evolution, with human nature itself being the Philosophers' Stone, to be transformed from its current "base metal" status to the "pure gold" of something transcending its own biology.

a. A Form of Magical Reversal of the Tower of Babel Moment

Ray Kurzweil, one of the leading thinkers exploring these possibilities, minced no words when he observed that the technologies would ultimately merge "our biological thinking with the nonbiological intelligence we are creating,"[8] that is, that human consciousness itself is alchemically transformed. As Kurzwel observes, the possibility exists that our non-biological intelligence component will eventually be trillions of times more powerful than our biological component.[9] Kurzweil also minces no words on the magical nature of modern science, and specifically, the ability of computers to game various scenarios and modify our environment:

A word on magic: when I was reading the Tom Swift Jr. books, I was also an avid magician. I enjoyed the delight of my audiences in experiencing apparently impossible transformations of reality. In my teen years, I replaced my parlor magic with technology projects. I discovered that unlike mere tricks, technology does not lose its transcendent power when its secrets are revealed. I am often reminded of Arthur C. Clarke's third law, that "any sufficiently advanced technology is indistinguishable from magic."

Consider J.K. Rowling's Harry Potter stories from this perspective. These tales may be imaginary, but they are not unreasonable visions of our world as it will exist only a few decades from now. Essentially all of the Potter "magic" will be realized through... technologies.... Playing quidditch and transforming people and objects into other forms will be feasible in full-immersion virtual reality environments, as well as in real reality, using nanoscale devices. More dubious is the time reversal (as described in *Harry Potter and the Prisoner of Azkaban*), although serious proposals have been put forward for accomplishing something along these lines (without giving rise to casulaity paradoxes), at least for bits of information, which essentially is what we comprise...

Consider that Harry unleashes his magic by uttering the right incantation. Of course, discovering and applying these incantations are no simple matters. Harry and his colleagues need to get the sequence, procedures, and emphasis exactly correct. the process is precisely our experience with technology. Our incantations are the formulas and algorithms underlying our modern-day magic. With just the right sequence, we can get a computer to read a book out loud, understand human speech, anticipate (and prevent) a heart attack, or predict the movement of a stock-market holding. If an incantation is just slightly off mark, the magic is greatly weakened or does not work at all.[10]

This technological transformation of mankind will, in turn, transform "every institution and aspect of human life, from sexuality to spirituality."[11] Mortality itself "will be in our own hands" for the technologies will allow one to live "as long as we want (a subtly different statement from saying we will live forever.)"[12] In other words, the ultimate goal of transhumanism is nothing less than the scientific and technological reversal of the Tower of Babel Moment of History, of the Fall of Man, and the alchemical ascent back up the *scala caeli*, the ladder to heaven.

b. Man the Microcosm Becomes the Macrocosm

An inventory of these "keys to creation," the "rungs" on the ladder ascending back to heaven, so to speak, is breathtaking. At the head of the list is the alchemical fusion of man and mineral itself, in the form of implanted computer chips to "enhance" the abilities of humans to interface directly with computers. Already Duke University has a "telekinetic monkey" named Belle, a small "owl monkey" with a computer implant in her brain, allowing her to move a mechanical arm hundreds of miles away from her actual location, simply by "thinking" it. The implants themselves are the latest in sophistication, being probes much finer than sewing threads, aligned with individual neurons with the region of her brain controlling motor skills.[13]

The implications of this are, in turn, equally breathtaking, for it means that any human being - a microcosm - will be capable of himself or herself becoming a macrocosm, of literally stretching out, via computer interfaces, to control robotic space probes, millions of miles away. The next step is towards true computer-enhanced telepathy and interface, "to rig a distant machine such that it can pipe what it is sensing directly into the brain of its human host. The goal is to seamlessly merge mind and machine, engineering human evolution so *as to directly project and amplify the power of our thoughts throughout the universe.*"[14]

c. Downloading and Uploading Memories, and Direct Modification of Consciousness and Behavior

All of this points out implicit promises, and dangers. The promise is obvious enough: people with weakened of lost motor skills would, through this new kind of prosthesis, be able to control artificial limbs. Or take the hippocampus in the brain, the structure that is crucial for processing and long-term memory.[15] Neurophysiologists at the University of Southern California are currently attempting to map the signals structure of this crucial region of the human brain. So successful have they been, they have developed a mathematical model "of the transformations performed by layers of the hippocampus and programmed the model onto a chip."[16] Additionally, these scientists have "a plan," which is to test the chip in animals by "disabling" their hippocampus, recording the loss of memory, and then by implanting a chip, to see if "that mental function can be restored..."[17] If successful, this technology would do an end run around epilepsy, Alzheimer's disease. It is, as Kurzweil aptly observes, a kind of "reverse engineering" of individual human brain.[18] Indeed, as he points out, the Max Planck Institute for Human Cognitive and Brain Sciences, based in Munich, has already engineering a computer chip that allows human neurons to be grown on a computer chip, interfacing *directly* with it.[19] The same institute has also developed technologies that allow it to detect when *specific* neurons are fired, and this same technology also allows them to *cause or prevent* certain neurons from firing,[20] in short, to modify human consciousness and behavior *directly* via computer implants.

Education itself will change profoundly once these technologies are widely available, for such direct downloading of information means that individuals will be able to disseminate and acquire information much more quickly.[21] This, coupled with the extension of human life itself, will mean that human knowledge will grow exponentially *within a given generation and individual*; we will be approaching the state ascribed to the gods in the ancient texts. Schools, if they exist at all, may consist of "downloading stations," and the typical school day may consist of no more than "five minutes" of "downloading" what now takes an entire day of lecturing to do.

But the dangers are equally if not more breathtaking, for one could program almost any "memory" into such a chip, and thereby remake an individual's personality, memories, and gain a new level of influence over their behavior. This opens up other possibilities for the emerging "alchemo-mineral" man. Kurzweil sees the possibility that entire individual brains could be "scanned" to "upload" an individual's personality and memories into a "new body" once the old one wore out, giving rise to the possibility of extreme biological longevity for a particular individual.[22] But why bother? With the emerging genetic technologies, replacement organs unique to the individual could be "grown" for harvesting and transplant when the old ones wore out.[23] Nano-machines would also contribute to life-extension simply by repairing diseased cells or tissue on a cell-by-cell basis.

d. The Group Consciousness? or Roddenberry's "Borg"?

Kurzweil points out yet another possibility that will emerge with the new technologies of direct neurophysiological manipulation:

"Experience beamers" will send the entire flow of their sensory experiences as well as the neurological correlates of their emotional reactions out onto the Web, just as people today beam their bedroom images from their Web cams. A popular pastime will be to plug into someone else's sensory-emotional beam and experience what it's like to be that person.... There will also be a vast selection of archived experiences to choose from, with virtual experience design another new art form.

... The most important application of circa-2030 nanobots will be literally to expand our minds through the merger of biological and non-biological intelligence. The first stage will be to augment our hundred trillion very slow interneuronal connections with high-speed virtual connections via nanorobot communication. This will provide us with the opportunity to greatly boost out pattern-recognition abilities, memories, and overall thinking capacity, as well as to directly interface with powerful forms of nonbiological intelligence. The technology will also provide wireless communication from one brain to another.[24]

In other words, the world-wide web will become exactly the kind of "super-brain" or "super-consciousness" long ago envisioned by Nikola Tesla. This highlights yet another danger, for how does one maintain one's individual personality and freedom in such a world, given the ability of these technologies to directly modify memory, behavior, and individual consciousness? And who will be the "systems administrators" or "operators" in this brave new world? What is being engineered, in other words, is literally a technological version of the corporate person, where it is no longer a legal metaphor, but a technological reality,[25] a single "distributed and interconnected brain," according to a study of the U.S. National Science Foundation and The U.S. Department of Commerce.[26] As we shall see momentraily, it is precisely such questions that have led some transhumanists to question whether the coming brave new world is in fact the heaven that others maintain it to be, for what is really being advocated is the alchemical version of a real, genuine telepathy, via technology.[27]

2. DARPA's Quest for the Transhumanist Supersoldier

As will be evident, all of the aforementioned technological possibilities have potential military applications, and no government agency on Earth is more acutely aware of this than is the Defense Advance Research Projects Agency, or DARPA. Briefly stated, transhumanist potentials also represent the possibility of supersoldiers. This quest - at least in modern times - began in part with the arms race spurred by the Cold War, and with the goal of try-ing to do a technological end run around thermonuclear weapons, to make wars between the major powers "fightable" and "winnable" again.

To gain an entry into the sort of bizarre "out-of-the-box" thinking that DARPA normally engages in, consider only whales or dolphins, and the fol-lowing chain of reasoning:

1) Whales and dolphins are sea-dwelling mammals;
2) As such, they can never sleep, at least in the conventional sense, for if they "slept" as land-dwelling mammals do, they would drown;
3) Therefore their neurophysiological processes are different and possibly of value to create a human soldier that would need far less sleep, or no sleep at all;
4) Research has shown that when whales and dolphins are "sleep-ing," one half of their brain is awake and working, while the other half is asleep. And this leads to the question, and the conclusion, that DARPA drew: "What would happen if humans could con-trol which portion of their brain is working while another por-tion recharges?"[28] DARPA's Continuous Assisted Performance program's, or CAP's, mission brief is to discover what happens, and how to do it.[29]

DARPA's CAP program has already conjectured that this single transfor-mation of mankind would lead to fundamental changes in the pace, tim-ing, and length of military operations;[30] it would be, in terms of military parlance, a "force multiplier" especially against an enemy that did not have the benefit of such technologies. Or suppose that through genetic enhance-ments, or nanotechnology, one breath of oxygen would permit a soldier to run for fifteen minutes, or carry his own weight in a backpack?[31] Or imagine x-ray goggles, like Superman, or exoskeletons to enhance strength and allow a soldier to "leap a tall building in a single bound." These are not the comic-book fantasies of yesteryear, but technologies actively being investigated by DARPA *now*.[32] Projects are already underway to design an exoskeleton, with a computer program reading an individual's muscle movements, and transmitting these to the exoskeleton which greatly amplifies the movements. Moreover, if the exoskeleton were immediately directed by the brain-computer interfaces referred to previously, the response of the exoskeleton would be as virtually instantaneous as the human body's response to the brain.[33] It is yet another example of the projected superpowers of transhumanist alchemo-mineral man.

Indeed, one DARPA projects manager and "future-predictor" summed up the quest for the transhumanist super-soldier in no uncertain terms:

> Soldiers having no physical, physiological, or cognitive limitations will be key to survival and operational dominance in the future.... Indeed, imagine if soldiers could communicate by thought alone.... Imagine the threat of biological attack being inconsequential. And contemplate, for a moment, a world in which learning is as easy as eating, and the replacement of damaged body parts as convenient as a fast-food drive-through. As impossible as these visions sound or as difficult you might think the task would be, these visions are the everyday work of the Defense Science Office. The Defense Sciences Office is about making dreams into reality,,,, These bold visions and amazing achievements... have the potential to profoundly alter our world.... It is important to remember that we are talking about science action, not science fiction.[34]

More about the Defense Sciences Office in a moment.

a. Longevity

For the present, let us concentrate on the implication of soldiers immune to any sort of biological threat - say a particularly virulent strain of *e coli*, or a genetic weapon targeting a certain race, or even a genetically engineered virus such as AIDS, engineered to accomplish in mere hours what normally is accomplish in years. DARPA's Continuous Assisted Performance or CAP program is also seeking that common ingredient to *all* pathogens, and thereby, seeking a "universal antidote" that can interrupt all of them. In short, DARPA is seeking a sort of "universal vaccine" against any and all biological agents.[35] If this sounds familiar, that's because it is, for this "universal antidote" are precisely the claims for the healing properties of the alchemical Philosophers' Stone, which, in its elixir form, was said to be able to heal virtually any disease, and to convey longevity.[36]

b. DARPA's and the Defense Sciences Office's Mission Briefs

But what is the Defense Sciences Office(DSO)? Quite literally, the DSO is a department within DARPA dealing specifically with realizing the transhumanist agenda in order make the transhumanist supersoldier a reality. It is, as its own members state, "DARPA's DARPA," the "cutting edge of the cutting edge."[37] It is worth pausing here to consider a possible historical antecedent to DARPA. Joel Garreau aptly summarizes DARPA's "mission brief" in the following fashion:

> A project is regarded as "DARPA-esque" only if few others would tackle it, but it would be earth-jolting if it did work. If you don't have falures, you're not far enough out. DARPA managers view themselves as instigators. By the time something new is mainstream enough to attract academic conferences attended by several hundred researchers, (the Defense Sciences Office) usually sees its midwife work as done and moves on to new challenges.[38]

The mission brief of the DSO, and for that matter, of DARPA, is *social engineering*, i.e., to engender enough interest in a particular research topic that a critical mass of human researchers is involved in the project to allow it to come to eventual fruition. Its role, in other words, is as a coordinating think tank, much like the *Kammlerstab* of Nazi SS general Hans Kammler during World War Two. Like General Kammler's *Kammlerstab*, DARPA's chief role is as a projects coordinator, to create a secret community of interaction between scientists secretly investigating radical concepts.[39] It is, in some respects, a breakaway civilization, and the projects it envisions will, as we shall also discuss below, will lead to a serious schism, to a breakaway civilization, within human society.

c. Back to Longevity

Certain organisms - tadpoles, starfish, even some lizards - have the ability to regrow a tail or limb that has been lost for whatever reason. Some DARPA denizens maintain that humanity itself once had, then lost, this ability.[40] But whether it did, or did not, is from one point of view, immaterial, since genetic splicing techniques could conceivably isolate the particular sequence in starfish, for example, which could then be spliced into other organisms - humans - which would allow them to regrow or repair severed limbs or wounds. And that opens up the possibility, as we have seen, of a kind of "virtual immortality," thus blurring the ancient mythological distinction between man and gods,[41] and, from a certain point of view, returning mankind to the divine status from which he descended according to certain ancient texts. Some go so far as to seriously propose lifespans in line with Old Testament texts, spanning hundreds of years, if not an entire millennium;[42] the GRIN technologies, in other words, are making the alchemical Philosophers' Stone, with its curative and longevity-bestowing properties, possible.

For Ray Kurzweil, the promise of nanotechnology "will ultimately enable us to redesign and rebuild, molecule by molecule, our bodies and brains and the world with which we interact"[43] thus leading to the same inevitable conclusion: longevity and virtual immortality. This longevity, plus the linked technologies interfacing human biological intelligence with the interconnected non-biological intelligence of computers, and via this interface, to other humans, opens up - at least as far as Kurzweil is concerned, the ability to socially engineer human society and institutions on a sweeping scale, in a true alchemical transformation of mankind, from his interior conscious and emotional life, to its outward manifestations in society and civilization.[44] As longevity increases, and as interconnected human knowledge increases exponentially, the distinction between work and play will disappear.[45]

But not everyone studying the alchemical transhumanist transformation of humanity is convinced that the coming longevity will issue in the "inevitable" increase and expansion of human knowledge. It was the physicist Max Planck who cynically observed that scientific paradigms change and progress when enough of the "old guard" dies off, and a new generation, more open to new ideas, takes the reins. But with longevity, that prospect fades, "This is," as Joel Garreau observes, "an intriguing hypothesis." [46] If knowledge stagnates with longevity, the constantly exponentially increasing curve of human progress levels off as an unintended consequence of the Singularity itself.

However, there is another possibility, one that Joseph mentioned in his very first book, and it is based upon the well-known fact that all technologies come pre-packaged with moral and ethical implications, for the kind of longevity being predicted by the transhumanists and which is just beginning to "come online"

...would mean one of two things for an individual in such a society. Either it would permit great moral progress in and toward the good, or great moral decay and "progress" in evil. Imagine an Albert Schweitzer or a Mother Teresa having thousands of years to do their work, or, conversely, an Adolf Hitler or a Joseph Stalin, and one has a picture of the moral condition such a society might be in.[47]

Indeed, the bright and sunny future scenarios envisioned by some transhumanist alchemists is not the only possibility. There are other scenarios being gamed and predicted.

B. The Scenarios of the Transhumanist Apocalypse Theater
1. The Three Scenarios

There are actually three broad possibilities as transhumanists contemplate the future and the human society that emerges from the new technologies:[48]

1) The "Heaven" scenario, in which the emerging technologies portend a benign and blissful future of longevity, of work as play, of sweeping extensions of human group consciousness not only globally, but on a cosmic scale;[49]
2) The "Hell" scenario, in which the same technologies lead humanity to a catastrophic end, due to accidental leaks of deadly viruses, or cataclysmic wars utilizing the new weapons, or through simple inability of human society and humans to cope with the sweeping changes as old institutions break down under the technological weight, and crumble into anarchy;
3) The "Prevail" scenario is, as might be expected, a mixture of the previous two, full of promise, to be sure, but also of reverses and set-backs, until ultimately, humanity makes choices regarding the technologies, and "muddles thought;"

2. The Assumptions of the Scenarios of the Transhumanist Apocalypse Theater

All the versions of the transhumanist scenarios, however, have, as Garreau notes, certain common features, and for our purposes, these are quite important, for they highlight as nothing else does, the alchemical nature of their underlying assumptions:

1) All predictive scenarios, no matter how elegantly programmed into super-computers, are based on assumptions about the future, and to that degree, are faith-based;[50]
2) All versions of the scenario *view human nature itself as an open, evolving system,*[51] in short, human nature is a transformative, information-creating medium, and to that extent, as we shall see in a later chapter, it is a mirror image of the physical medium itself,

and as such, is deeply and intimately connected to it. In short, human nature, as a transmutative information-creating medium, is itself the alchemical "Philosophers' Stone". It is this feature of human nature that allows the "engineered evolution"[52] that is the central component to each transhumanist scenario;

3) All scenarios have certain limiting rules or defining parameters:

a) The first, and most obvious, is that "they must conform to all known facts;"[53]

b) All scenarios must identify those predetermined factors that lock future events into present or past developments such that the future is more or less predetermined;[54]

c) All scenarios must determine, identify, and adequately model "critical uncertainties," or those logical possibilities that might occur within a margin of probability, and that are highly important if they *do* occur;[55]

d) All scenarios must attempt to identify highly improbable occurrences, or "wild cards", that would have tremendous impact if they did occur;[56]

e) Scenarios should attempt to identify their "embedded as-sumptions - such as the points enumerated in this inventory - so as to be able to determine the indicators or "early warn-ing signals" that a particular scenario is coming to pass;[57]

4) All scenarios are subject to certain types of predictive failures:

a) The situation was more complicated than that originally modeled;[58]

b) The cost-to-benefit ratio was inadequately modeled, or not even considered;[59]

c) The projected future did not adequately consider new tech-nologies, and was overtaken by them,[60] a significant diffi-culty, since all transhumanist scenarios are precisely about future transformations by technology;

d) Some adverse experience with a particular technology pre-vented its rapid expansion, for example, nuclear disasters such as Fukushima or Three Mile Island have inhibited the spread of nuclear fission plants;[61]

e) Predictive scenarios that do not adequately consider human behavior. Garreau cites the example of confident predictions, a few years ago, of "paperless offices," as computers would eliminate the need for shuffling forms.[62]

As is evident from this list, each of the three versions of the "Singularity" - the Heaven, Hell, and Prevail scenarios - are all predicated on one common assumption, in addition to those enumerated above, and that is, that one of the three will inevitably occur *provided that some cataclysm does not disrupt the curve of exponentially increasing technology*, a cataclysm such as a killer asteroid striking the Earth, or a nuclear or biological war.[63]

With respect to the two scenarios at the opposite ends of the spectrum, Garreau notes that among the signs that humanity is entering the "Heaven" scenario are an increase of unimaginably good things occurring, such as the conqeust of poverty, diseases, or even an increasing popular use of transhumanist terminology, such as the term "Singularity" itself, much like "global warming" became part of the popular conceptual vocabulary at the end of the 20th and the beginning of the 21st centuries.[64] At the opposite end, the Hell scenario, the warning indicators or its beginning phases would be that the growth of complexity begins to slow, or becomes erratic, unimaginably bad things start to happen as new diseases become rampant, or ideological or religious systems gain control over the advance of technologies and prevent their development,[65] or weapons of mass destruction based on the GRIN technologies rapidly proliferate.[66] In the middle, the Prevail scenario would be indicated by early warning signs such as researchers voluntarily ceasing work on various research projects considered too dangerous, or declining funding from corporations or military agencies with dubious histories, or ceasing work on projects deemed "too fraught with human peril."[67]

We leave it to the reader to decide which, if any, of these transhumanist apocalypse theater plays we are watching. But one thing should now be evident: each scenario, as Joel Garreau so aptly put it, depends upon the Promethean ambition of "stealing fire from the gods, breathing life into inert matter and gaining immortality."[68] But this Promethean ambition, to literally "unbind" Prometheus in the alchemical engineering of "enhanced humans," contains yet another hidden danger, namely,

C. . THE POSSIBILITIES AND DANGERS OF A BREAKAWAY CIVILIZATION: "ENHANCED" VS. "NORMAL" HUMANS.

Let us imagine a society in which the two kinds of humans, the alchemically-enhanced and new-fangled engineered man, stuffed to the gills with nanorobots repairing his cells, with genetically modified DNA to make him disease and aging-resistant, and chipped to plug into the world-wide-web and into other similarly enhanced humans, and "the rest of us" normal, non-modified man. The difference between the two will literally constitute the former engineered version of humans into a new elite, a civilization unto itself, for as Garreau notes, this new humanity will be able to

1) Think faster and more creatively;
2) Possess remarkable and nearly photographic memories;
3) Read entire books with almost total comprehension and recall, in mere minutes;
4) Live long, and even enbark on several *different* careers;
5) Repair their bodies without having to visit physicians, and maintain peak health without exercise;
6) go with a minimum of sleep;
7) communicate with each other nearly instantaneously, allowing coordinated action to occur much faster than among "normals."[69]

As the Singularity approaches, in other words, there will be a transitional period of time when there are literally two humanities, and two types of human civilization, one approaching Promethean, godlike capabilities.

Some advocates of the "Hell" scenario contend that the "alchemo-mineral" transformation of mankind will lead to the ultimate Malthusian nightmare, as longevity greatly overpopulates the earth. Ray Kurzweil disagrees, pointing out that the GRIN technologies will also lead to a vast expansion of wealth creation.[70] But again, this radical wealth creation would initially remain in the hands of the very wealthy who are the ones most readily positioned to take advantage of the new GRIN technologies as they become available, thus creating yet another condition for a "breakaway civilization" to emerge.

For Kurzweil, and proponents of the Heaven scenario, however, the concept of the breakaway civilization is the actual *cosmic* goal as man the microcosm becomes man the macrocosm in reality, not just myth:

In the aftermath of the Singularity, intelligence, derived from its biological origins in human brains and its technological origins in human ingenuity, will begin to saturate the matter and energy in its midst. It will achieve this by reorganizing matter and energy to provide an optimal level of computation... to spread out from its origin on Earth.

.... In any event the "dumb" matter and mechanisms of the universe will be transformed into exquisitely sublime forms of intelligence, which will constitute the sixth epoch in the evolution of patterns of information.

This is the ultimate destiny of the Singularity and of the universe.[71]

In this view, the "Heaven" transhumanists have emphasized the information-creating nature of modern physics,[72] as we shall discover in the final chapter. Indeed, these transhumanists, taking a cue from Konrad Zuse, the inventor of the first digital computers for the Nazis,[73] who first proposed the idea that the entire universe was a digital algorithm being run on a computer.[74] By thus viewing the universe as a "giant cellular-automaton computer" some physicists are entertaining the notion that "apparently analog phenomena (such as motion and time)" as well as the actual physical formulae themselves can be modeled as "simple transformations of a cellular automaton" in almost Chomskian fashion, as a huge "transformational-generative grammar," an algorithm.[75]

The essence of the transhumanist vision, then, ultimately depends on models of physics that we will examine in the final chapter, for that vision view the universe and the underlying physical medium from which it arises in genuinely alchemical fashion, as a Philosophers' Stone.

In their contemplations of the creation of GRIN technologies-modifed "alchemo-mineral" man, the transhumanists are in fact envisioning nothing but the old alchemical homunculus.

But there is one, final rung on the ladder of alchemical ascent that few, including many transhumanists, like to discuss, the most disconcerting image of them all: the primordial androgyny from which, in esoteric lore and doctrine, all else derives. It is an androgynous future portended, for example, in some research projects, such as the one at the University of Pennsylvania, where male mouse cells were genetically transformed into egg cells, allowing two male mice to become the parents of a third, with obvious implications for yet another alchemical transformation of humanity, an androgynous one.[76]

All of this, as we shall see in chapters eight and nine, was foreseen by theological and poetical prophets, and in some very unlikely places. But before we can begin to consider that final rung, that final most disconcerting "alchemosexual" image, we must summarize the results of our study thus far, to place it within the proper context.

ENDNOTES

1 Arthur C. Clarke, *Profiles of the Future*, 1961.

2 Percy Bysshe Shelley, *Prometheus Unbound*, Act I, 371-375, Donald H. Reiman and Neil Fraistat, eds., *Shelley's Poetry and Prose: Norton Critical Edition*, Second Edition (W. W. Norton and Company: 2002), p. 221.

3 Joel Garreau, *Radical Evolution: The Promise and Peril of Enhancing Our Minds, Our Bodies - And What It Means to Be Human* (Broadway: 2006), p. 11.

4 Ibid., p. 4. See also Ray Kurzweil, *The Singularity is Near: When Humans Transcend Biology* (Penguin, 2005), p. 205.

5 Garreau, op. cit., p. 4.

6 Kurzweil, op. cit., p. 205.

7 Garreau, op. cit., p. 6.

8 Kurzweil, op. cit., p. 4.

9 Ibid., p. 9.

10 Ibid., pp. 4-5.

11 Ibid., p. 7.

12 Ibid., p. 9.

13 Garreau, op. cit., pp. 19-20.

14 Ibid., p. 20, emphasis added.

15 Kurzweil, op. cit., p. 188.

16 Ibid.

17 Ibid.

18 Ibid., p. 4.

19 Ibid., p. 195.

20 Ibid., p. 313.

21 Ibid., p. 337.

22 Ibid., pp. 198-199.

23 Ibid., pp. 222-223.

24 Ibid., p. 316.

25 Garreau, op. cit, p. 75.

26 Ibid., p. 114, citing the report *Converging Technologies for IMproving Human Performance*. Garreau also refers to Gregory Stock's books, whose titles say it all: *Redesigning Humans: Our Inevitable Genetic Future*, and *Metaman: The Merging of Humans and Machines into a Global Superorganism*. The consciousness implications of such mergers were major themes of science fiction author Isaac Asimov in his *Foundation* series of novels.

27 Ibid., p. 37.

28 Ibid.,p. 28.

29 Ibid.

30 Ibid.

31 Ibid., p. 4.

32 Ibid., pp. 5, 20-21.

33 Ibid., p. 36.

34 Ibid.,pp. 22-22.

35 Ibid., pp. 29-30.

36 See Joseph P. Farrell, *The Philosophers' Stone: Alchemy and the Secret Research for Exotic Matter* (Feral House, 2009), pp. 76-78.

37 Garreau, op. cit., p. 25.

38 Ibid., p. 41.

39 See Joseph P. Farrell, *Reich of the Black Sun: Nazi Secret Weapons and the Cold War Allied Legend* (Adventures Unlimited Press, 2004), pp. 104-107. It is important to note that the idea of the Black Sun is both an ancient Babylonian doctrine, with its

own alchemical references to the transformation of humanity, since it was a "secret sun," hence the name "Black Sun," of the illumination or transformation of human consciousness.

It is also worth noting that Garreau mentions that one area being investigated by DARPA's Information Processing Technology Office (IPTO) is a "special focus area" called "Time Reversalo Methods."(Garreau, op. cit., p. 41)

40 Garreau, op. cit., p. 28.
41 Ibid., p. 53.
42 Ibid., p. 91.
43 Kurzweil, op. cit., p. 227.
44 Ibid., p. 299.
45 Ibid., p. 300. See also p, 323.
46 Garreau, op. cit.,p. 163.
47 Joseph P. Farrell, *The Giza Death Star* (Adventures Unlimited Press, 2001), p. 289.
48 For a further discussion, see Garreau, op. cit., pp. 3-13.
49 It should be noted that Garreau actually outlines a fourth scenario, the "Transcend" scenario, on pp. 227-265. For our purposes, we view the Heaven and Transcend scenarios as two different versions of the same alchemical re-ascent up the animal, vegetable, mineral, and androgyny ladder.
50 Garreau, op. cit., pp. 208-209.
51 Ibid., p. 232.
52 Ibid., p. 231, for more on Garreau's elaboration of "engineered evolution."
53 Ibid., p. 78.
54 Ibid.
55 Ibid., p. 79.
56 Ibid.
57 Ibid.
58 Ibid., p. 211.
59 Ibid.
60 Ibid.,p. 212.
61 Ibid.
62 Ibid.
63 Ibid., p. 83.
64 Ibid., p. 130.
65 Ibid., p. 131.
66 Ibid., p. 139.
67 Ibid.,p. 225.
68 Ibid., p. 106.
69 Ibid., pp. 7-8.
70 Kurzweil, op. cit., p. 13.
71 Ibid., p. 21. See also pp. 28-29.
72 Ibid., p. 85.
73 For Konrad Zuse's little-known contribution to the development of digital comput- ing, see Joseph P. Farrell, *Reich of the Black Sun: Nazi Secret Weapons and the Cold War Allied Legend* (Adventures Unlimited Press, 2004), pp. 187-188.
74 Kurzweil, op. cit., p. 86.
75 Ibid.
76 Ibid., p. 12.

⚛ Seven ⚛

RECAPITULATION:
CONCLUSIONS THUS FAR

∴

"And Science struck the thrones of Earth and Heaven,
Which shook but fell not; and the harmonious mind
poured itself forth in all-prophetic song,
And music lifted up the listening sprit
Until it walked, exempt from mortal care,
Godlike, o'er the clear billows of sweet sound,
And human hand first mimicked and then mocked
With moulded limbs more lovely than its own
the human form, till marble grew divine..."
—Percy Bysshe Shelley, *Prometheus Unbound*, II.iv, 74-83.[1]

LET US PAUSE FOR A MOMENT, and take inventory of what we have uncovered thus far. In part one, we explored the Tower of Babel Moment and the Topological Metaphor of the medium, and, just to refresh our memories, these are the conclusions drawn from that section:

1) Its initial unity, which expresses itself in three primary ways:
   a) As a sexual, or androgynous unity;
   b) As a linguistic unity; and finally, and perhaps most importantly,
   c) As a cultural-philosophical unity. In this instance, the "Topological Metaphor" also reveals the fact that this ancient philosophy was exactly what the Mediaeval and Renaissance Hermeticists claimed it was, the *prisca theologia*, the "ancient theology."

2) This unity in all its facets constituted some sort of threat to the gods or God, and had to be broken. Notably, when one looks at *all* the ancient records, the unity was indeed broken *at each of the three levels noted above.*

Now, we may add to these conclusions, those of this section:

3) Modern science, particularly physics, has deep hermetic, al-chemical origins, and although the *methods and techniques* have changed, the *alchemical goals* for the transformation of mankind have not;

4) It was, in fact, alchemy itself and its claims to be able to engineer artificial life, that first raised the transhumanist prospect of the apocalyptic transformation of man;

5) In the transhumanists' hands, this transformation is ultimately envisioned as the transformation of the entire cosmos into the model of human consciousness, via implants and networked computer connections between humans, and the extension of human technology throughout the universe;

6) In this alchemical vision, all aspects of man and his environment, are to be transformed, from his very being to the food he eats;

7) Oddly enough, modern science appears to be ascending back up the ladder of the descent of man as understood by the esoteric tradition and by the ancient mythologies of the Tower of Babel Moment and Fall of Man, and to be doing so in the exact order and reversal of that descent, suggesting once again that the goals of modern science have remained, at root, alchemically inspired;

8) Into this picture, however, we also discovered that the Three Great Yahwisms have injected a new element, that of the trans-formation of alchemical principles into the social engineering of man, as the "regions" of the ancient topological metaphor were transmuted into regions of "believer and infidel" within human social and cultural space.

As we shall now discover, all these features combine to reveal an interesting program and agenda concerning the topmost rung on that ladder of descent, the primordial androgyny itself. It is precisely here, as we shall also discover, that old inherent contradictions between those monotheisms and science are most in evidence. And finally, it is here that we shall also discover a very hid-den, though for those able to read the symbolism, very palpable, androgynous symbolism within the Western esoteric tradition and the most prominent secret societies preserving it. It is here that we shall discover all the alchemical themes being openly paraded in some of the great literary giants of the nineteenth century.

ENDNOTES

1   Percy Bysshe Shelley, *Prometheus Unbound* in *Shelley's Poetry and Prose: Norton Critical Edition*, ed. Donald H. Reiman and Neil Fraistat (W.W. Norton & Company, 2002), p. 249.

III.

THE ANDROGYNOUS GOD OF ALCHEMY AND THE ALCHEMOSEXUAL ASCENT

"And the beasts, and the birds, and the insects were drowned
In an ocean of dreams without a sound;
Whose waves never mark, though they ever impress
The light sand which paves it, consciousness...."
—Percy Bysshe Shelley, *The Sensitive Plant,*
lines 102-105

"The one duty we owe to history is to re-write it."
—Oscar Wilde,
"The Critic as Artist"

"All art is at once surface and symbol.
Those who go beneath the surface do so at their own peril.
Those who read the symbol do so at their own peril."
—Oscar Wilde,
Preface to *The Picture of Dorian Gray*

$$\exists\varnothing_{x,y}:\P\varnothing_{x,y}\to\varnothing^0_{x,y}+\partial\varnothing_{x,y}+\varnothing^0_x$$
$$:(2)\varnothing^0_{x,y}\cap\varnothing^0_{x,y}\to\partial\varnothing_{x,y}$$
$$:(3)\varnothing^0_{x,y}\cap\varnothing^0_x\to\{\tfrac{\partial\varnothing_x}{\partial\varnothing_{x,y}}\}$$

where elements within brackets, {}, are Boolean "and/or" operators

≋ Eight ≋

"Aquinas," Alchemosexuality,
and the Androgynous God:
Transmutation, Transubstantiation, and Transitions

∴

*"Br. Thomas replied: 'I cannot do it, Reginald, everything I have written
seems as worthless as straw.'"*
—"Processus Canonizationis S. Thomae Aquinatis Neapoli"[1]

*"The essence of this sacrificial act
lies, first, in **what** one does, and, second, in **how** -
the **matter** and the **manner** of the pact."*
—Dante Alighieri, *The Divine Comedy: Paradiso*, Canto V, 53-45.[2]

BEYOND QUESTION, the most famous theologian-philosopher of the
High Middle Ages was Thomas Aquinas(1225-1274), a saint and doctor of
the Roman Catholic Church, whose influence upon that institution, and
therefore through it upon the world, was profound and long-lasting. It can,
indeed, be felt to this day, for many of the Roman Chuch's formulations of the
doctrines unique to it owe much to him, and his influence on philosophy and
theology spread far beyond its borders and can be felt down through history.
Additionally, Aquinas' influence can be felt beyond the confines of cathedral
or cloister, for it is a commonplace in literary criticism that he influenced
Dante Alighieri's *Divine Comedy*.

By the time Aquinas suddenly stopped writing on philosophy and theology altogether, he had poured out a veritable library of biblical commentary,
treatises on Aristotle, and two massive theological works on almost all aspects

of Catholic Christian doctrine. Therein lies the problem, for having devoted a lifetime to scholarship, research and writing on theology, Aquinas just suddenly *stopped* doing so altogether.

The question is, why?

The answer of all parties concerned is simple: he had a *vision*.

The real question that divides them is, a vision of *what*?

In the answer to *that* question, there lies a story, or rather, *two* stories, the official ecclesiastical one, and the not-so-well-known, unofficial, and baldly esoteric one. One might say that the latter story has been carefully, perhaps deliberately, buried. But in either case and whichever version of the story was true, it remained a "theology-stopping" vision.

A. Aquinas' Theology-Stopping Vision

The question of the nature of Aquinas' vision assumes some importance, for as we shall discover in the next section, there is a highly alchemosexual document - the *Aurora Consurgens* - that is attributed in some quarters to Aquinas, and it is that document that throws the whole nature of his theology-stopping vision into controversy.

The vision itself is fairly straightfoward, but its implications are not, for in the middle of writing his massive compendium of Catholic theology, the *Summa Theologica*,

> while working on the section on penance, that celebrated experience befell him which put an end to his literary activity and, a month later, to his life. The report of it is contained in the *Acta Bollandiana* and rests on the testimony of Bartholomew of Piperno, Thomas's best friend and confidant.
>
> "Moreover the same witness (Bathrolomew of Capua) said that, when the said Brother Thomas was saying Mass in the said chapel of St. Nicholas at Naples, he was smitten with a wonderful change, and after that Mass he neither wrote nor dictated anything more, but suspended his writing in the third part of the *Summa*, in the treatise on Penance. And when the same Br. Reginald saw that Br. Thomas had ceased to write, he said to him: Father, why have you put aside so great a work, you began for the praise of God and the enlightenment of the world? And the said Br. Thomas replied: *I cannot go on*. But the said Br. Reginald, fearing that he had fallen into madness as a result of too much study, kept on pressing the said Br. Thomas to go on with his writing, and likewise Br. Thomas replied: *I cannot do it Reginald,*

everything I have written seems as worthless as straw. Then Br. Reginald, overcome with surprise, so arranged matters that the said Br. Thomas went to visit his sister, the Countess of San Severino, of whom he was very fond; he hastened to her with great difficulty, and when he arrived, and the Countess came to meet him, he hardly spoke to her. Then the Countess, in a state of great fear, said to Br. Reginald: What is all this? Why is Br. Thomas all struck with a stupor, and hardly speaks to me? And Br. Reginald answered: He has been like this since about St. Nicholas's day, and since then he has not written anything. And the said Br. Reginald began to press the said Br. Thomas to tell him for what reason he refused to write and why he was stupefied like this. And after a great many pressing questions from Br. Reginald, Br. Thomas replied to the said Br. Reginald: *I adjure you by the living God Almighty and by your duty to our Order and by the love you have for me, that so long as I am alive you will never tell anyone what I am going to tell you.* And he went on: *Everything that I have written seems to me worthless in comparison with the things I have seen and which have been revealed to me.*[3]

This is curious behavior, for why would a known and revered theological master such as Aquinas wish to conceal what was revealed to him, in whatever vision, or flash of insight, he had?

One answer - the answer of traditional Christian piety - would be that he was too humble, and that there were no words to describe it. But still, surely something more could have been said, other than that everything he had spent his life writing was "worthless in comparison" to that vision or insight. The other possibility is, of course, that he had seen something, intuited something, or had an insight that was altogether of a different order than the Christian. Indeed, this would rationalize the vision, and the unwillingness to write on Christian topics ever again, more reasonably.

We are told by the same witness, however, that on his deathbed Aquinas made the obligatory Christian statements, which would seem to cast out the last possibility:

And the said witness said moreover, that when the said Br. Thomas began to be overcome with sickness in the said village... he besought with great devotion that he might be borne to the monastery of St. Mary at Fossanova: and so it was done. and when the said Br. Thomas entered the monastery weak and ill, he held on to a doorpost with his hand and said: This is my rest for ever and ever.... And he remained for several days in that monastery in his ill state with great patience and humility, and desired to receive the Body of our Saviour. And when that Body was brought to him, he genuflected, and with words of wondrous and long-drawn-out adoration and glorification he saluted and worshipped it, and before receiving the Body he said: I receive thee, price of my soul's redemption, I receive thee, viaticum of my pilgrimage, for love of whom I have studied and watched and laboured and preached and taught: never have I said aught against thee unless it were in ignorance: nor am I obstinate in my opinion, but if I have said aught ill, I leave it all to the correction of the Roman Church. And then he died, and was buried near the high altar of the church of that monastery, where there is a stream, from which a water-wheel takes up water, by which all that place is watered, as the witness himself has often and carefully observed.[4]

While this may tend to suggest that Aquinas' theology-stopping vision remained within the bounds of Catholic orthodoxy, Marie-Louise Von Franz, a Jungian psychologist and commentator on the *Aurora Consurgens* attributed to him, points out one disturbing possibility, namely, that Aquinas' words "on receiving the viaticum are rather strange, for he tells Christ that he has never said anything against him - which suggests that the psychological possibility did exist of his saying something unorthodox."[5]

This possibility of "saying something unorthodox" brings us directly to the heart of the matter: the *Aurora Consurgens* attributed to Aquinas, and to the possibility that it contains the contents of that theology-stopping vision.

B. The Alchemosexual Vision of the Aurora Consurgens and Its Implications
1. Attributions and Authenticity of the Aurora Consurgens

In qualifying the vision contained in the *Aurora Consurgens* as alchemosexual, we go beyond evaluations of its androgynous imagery, for as we shall discover, the vision contains deep, rich, alchemical imagery and practice.

Indeed, it is this very fact that has led those defending the standard Christian interpretation of Aquinas' vision to reject any connection between the clear, elegant, logical prose of the Mediaeval schoolman and the florid, ecstatic, stream-of-consciousness utterances of the *Aurora*. Nonetheless, the mediaeval record is also quite clear, for there were clear attributions of the work to Aquinas, "an attribution so surprising and, at first sight, so unlikely that hitherto it has never been taken seriously. This is due, among other things, to the fact that the importance of the treatise was not recognized before."[6] So bewilderingly *different,* bizarrely mystical, and alchemosexual are its contents that, indeed, "the treatise has about it an air of strangeness and loneliness which, it may be, touched and isolated the author himself."[7]

The text itself mentions one key person, which increases the likelihood of Aquinas' authorship somewhat, and that is the towering figure of Albertus Magnus, Thomas' mentor and master in theology, and a well-known alchemist,[8] who had formed such a strong bond with his pupil and disciple that when the latter died, it is reported that Albertus felt and knew that Thomas had passed away telepathically, "as often happens in the case of an intense relationship with mutual projection."[9]

Additionally, the text cites numerous alchemical sources well-known prior to and during the time of Aquinas, but no sources written subsequently to his life:[10] "All the traceable sources are early Latin treatises or translations of Arabic writings, none of which can be dated later than the middle of the thirteenth century."[11] None of the later popular writers on alchemy such as Arnaldus de Villanova (1235-1313) or Raymond Lully (12-35-1315) are ever mentioned in the *Aurora Consurgens.*[12]

2. Style of the Aurora Consurgens, and a Problem

The *Aurora Consurgens* is an extended exposition and commentary on the Song of Solomon, a popular subject for Christian mystics... and for alchemists as well, though as can be imagined, they treat the subject very differently, and while the author of the *Aurora* struggles mightily to keep his ecstatic vision within the bounds of biblical imagery and Christian doctrine, by the end of the work, in the "Seventh Parable," the ecstasy has overwhelmed him, and the alchemosexual image of God and man have both overpowered any attempt to remain within the confines of mediaeval Christian orthodoxy. The psychologist Marie-Louise Von Franz, whose extensive commentary on the *Aurora* forms the argument and basis of our own interpretation here, has this to say about the enraptured style of the work, so unlike Aquinas' other works:

> In the last parable, which is largely a paraphrase of (the Song of Solomon), it rises to the point of ecstasy. It is difficult to withstand the impression that the whole treatise was composed in an abnormal psychic state. Moreover, minor inaccuracies in the quotations make it evident that these were reproduced on the spot and written down quickly. We can therefore conclude that the treatise must have been composed under unusual circumstances. The abnormal state of

mind seems to consist mainly in the steady flow of imagery, which guided the author's pen in a way that is ordinarily observed only during periods of intense excitement bordering on rapture or possession, when unconscious contents overwhelm the conscious mind. The loss of conscious control would explain the extraordinary manner of speech and expression to which the author had involuntarily to submit. *Aurora* is unique, not only with respect to the mystical literature of the time, but also among the authentic alchemical treatises of the period.[13]

But there is another massive problem with its style and contents, a problem clearly pointing to *someone* with extraordinary knowledge both of alchemical texts, of the biblical text, and of the texts of the mediaeval Catholic ritual:

> One is forced to the conclusion that it cannot have been written by an alchemist who lived entirely in the world of "chemical" ideas. Evidence for this is the fact that only about a dozen of the "classics" of alchemy are quoted, and only the most general sayings at that, while all evidence of any detailed knowledge of the material, as well as chemical recipes and technical instructions, together with the word "alchemia" itself, are lacking. In the case of a man who was only an alchemist, mention of these things would be almost obligatory. On the other hand, *we have to postulate for our author a fairly good acquaintance with the Bible and the liturgy. These facts combine to suggest that he was, above all, a cleric.* His praise of the "parylui" and "pauperes" might be an indication that he was a member of the Dominican or Franciscan Order.[14]

Aquinas was, of course, a Dominican, and a cleric.

This is not the only problem the *Aurora* poses that suggest a strong connection between it and Aquinas, for within the manuscript tradition of that text, at least three texts make no mention of the authorship - the Paris, Vienna, and Venice versions - while the Bologna and Leiden manuscripts explicitly ascribe the work to Aquinas.[15] The fact that the mediaeval manuscript sources are so totally and almost equally divided over the matter of the authorship of the *Aurora* suggests that its contents posed quite a problem for the ecclesiastical authorities.

All this puts the most puzzling feature about Thomas' output into stark relief, for it is well-known that Thomas preached a series of short sermons and commentary on the Song of Solomon, a work that has never come to light, and the witnesses who remark that he did indeed do this, are mysteriously silent as to its contents. The only manuscript upon that subject, and that has been attributed to Aquinas - and debated ever since - is precisely the *Aurora Consurgens*.[16]

3. The Seventh Parable and the Alchemosexual Androgyny of God and Man

If the silence of these witnesses was deliberate, then it is at least understandable, for by the time one reaches the end of the *Aurora* and its Seventh Parable "Of the Confabulation of the Lover with the Beloved," one has moved quite beyond the bounds of the Christian orthodoxy of that - or any other - period. The ecstatic prose reaches a fever pitch throughout the parable, until, at the very end, the author is so enraptured he speaks both in the voice of the Lover *and* the Beloved, from the alchemical conjunction of opposites, until a startling image bursts forth:

> The author of the *Aurora*, who no doubt was familiar with the Church's interpretation (of the Song of Solomon), employs (the Son) for the grater glorification of *his* (conjunction). As in the Biblical version, the speakers alternate, and the transition from one to the other is sometimes not at all clear. It is as if the two figures were speaking directly out of a state of non-differentiation...[17]

Then, the voice shifts to the author's soul, speaking at the very beginning of the Seventh Parable:

> Be turned to me with all your heart and do not cast me aside because I am black and swarthy, because the sun hath changed my colour and the waters have covered my face and the earth hath been polluted and defiled in my works.[18]

But the problem in the context of the quotation from the book of Joel is that the author abandons the biblical text, where God is speaking, and continues to speak of himself, and thus is speaking in a non-differentiated state, simultaneously as God and as Man, and speaking alchemically at that:

> Here the bride, or the prima materia or its soul, is speaking, and she begs for help and deliverance. but, as the quotation from Joel shows, she is at the same time identical with God. This is one of the places where that equation... is expressed most clearly: where God, or at any

rate his feminine aspect, appears as the spirit or soul in matter and awaits redemption by the work of man.[19]

In other words, in the ecstatic utterance of whatever vision the author of the *Aurora* was having or was trying to communicate, the distinction between God and Man, so crucial to Christian theology - the distinction between God as the Lover and the human soul as Beloved - has altogether collapsed. One is no longer moving in a Christian world at all, but in a world of the "chemical wedding" or conjunction, the fusion, of "opposites."

It gets worse (at least, from the Christian point of view), for the Seventh Parable only increases in this intense communion and ecstasy, so much so that

> The transition from one speaker to the other is barely perceptible, so that one is inclined to think that it is the same figure speaking now as a woman and now as a man, just as so often before (the biblical figure of) Wisdom coalesced with the Holy Spirit, Christ, or God. One has the impression that the author, having achieved direct contact with the unconscious, is letting the voices speak just as he hears them, without bringing his ego into it at all, as if he and the unconscious had again become identical.[20]

Then comes a lengthy passage in the Seventh Parable, constituting the bulk of the parable itself, so rich with a flooding ecstatic riot of biblical and alchemical imagery, that we reproduce it in its entirely below, before citing Von Franz's commentary which unpacks the flood of alchemosexual allusions:

> ...I will come forth as a bridegroom out of his bride-chamber, for thou shalt adorn me round about with shining and glittering gems and shalt clothe me with the garments of salvation and joy to overthrow the nations and all mine enemies, and shalt adorn me with a crown of gold engraved with the sign of holiness and shalt clothe me with a robe of righteousness and shalt betroth me to thee with thy ring and cloth my feet in sandals of gold. All this shall my perfect beloved do, exceeding beautiful and comely in her delights, for the daughters of Sion saw her and the queens and concubines praised her. O queen of the heights, arise, make haste, (my love), my spouse, speak (beloved) to thy lover, who and of what kind and how great thou art, for Sion's sake thou shalt not hold thy peace and for the sake of Jerusalem thou shalt not rest from speaking to me, for thy beloved heareth (thee)." "Hear all ye nations, give ear

all ye inhabitants of the world; my beloved, who is ruddy, hath spoken to me, he hath sought and besought. I am the flower of the field and the lily of the valleys. *I am the mother of fair love* and (of fear an) of knowledge and of holy hope. As the fruitful vine I have brought forth a pleasant odour, and my flowers are the fruit of honour and riches. I am the bed of my beloved, which threescore of the most valiant ones surrounded, all holding swords upon their thigh because of fears in the night. I am all fair and there is no spot in me; looking through the windows, looking through the lattices of my beloved, wounding his heart with one of my eyes and with one hair of my neck. I am the sweet smell of ointments giving an odour above all aromatical spices and like unto cinnamon and balsam and chosen myrrh. I am the most prudent virgin coming forth as the Dawn, shining exceedingly, elect as the sun, fair as the moon, besides what is hid within. I am exalted as a cedar and a cyprus tree on Mount Sion.... *wherefore have all the philosophers commended me and sowed in me their gold and silver and incombustible grain. and unless that grain falling into me die, itself shall remain alone, but if it die it bringeth forth threefold fruit: for the first it shall bring forth shall be good because it was sown in good earth, namely of pearls; the second likewise good because it was sown in better earth, namely of leaves (silver)l the third shall bring forth a thousand-fold because it was sown in the best earth, namely of gold....*I give and I take not back, I feed and fail not, I make secure and fear not; what more shall I say to my beloved? I am the mediatrix of the elements, making one to agree with another; that which is warm I make cold, and the reverse; that which is dry I make moist; and the reverse; that which is hard I soften, and the reverse. I am the end and my beloved is the beginning. I am the whole work and all science is hidden in me. I am the law in the priest and the word in the prophet and counsel in the wise.... I stretch forth my mouth to my beloved and he presseth his to me; *he and I are one; who shall separate us from love?* None and no man, for our love is strong as death.....Come my beloved and let us go into thy field...let us therefore enjoy them and use the good things speedily as in youth, let us fill ourselves with costly win and ointments, and let no flower pass by us save we crown ourselves therewith, first with lilies, then with roses, before they be withered. let no meadow escape our riot. Let none of us go without his part in our luxury, let us leave everywhere tokens of joy, for this is our portion, that we should live in the union of love with joy and merriment, saying: Behold how good and pleasant it is for two to dwell together in unity. Let us make therefore three tabernacles, one for thee, a second for me, and a third for our sons, for a threefold cord is not easily broken. He that hath ears to hear, let him hear what the spirit of the doctrine saith to the sons of the discipline concerning the espousal of the lover to the beloved. For he had sowed his seed, that there might ripen thereof threefold fruit, which the author of the Three Words saith to be three precious words, wherein is hidden all the science, which is to be given to the pious, that is to the poor, from the first man to the last.[21]

To the alchemically-knowledgeable theologian-scholars of the Middle Ages, this riotous ecstasy - and all the implications thereof - would have been immediately evident.

But to a modern reader, it will require the alchemical analysis of Marie-Louise Von Franz to make clear. The first thing to be noted about the passage is its reversal of the standard Christian imagery of the *Church* as the *bride* awaiting resurrection; but in the *Aurora*, it is the *bridegroom* - in Christian terms, Christ himself - who is awaiting resurrection. This is not, as we shall discover, the only disquieting case where the two have become blended, and then reversed.

Von Franz, referring to the previous parables in the *Aurora*, notes that

...(The) risen bridegroom stands at the left hand of the Queen, who appears to him "in gilded clothing, surrounded with variety." This figure, the Queen of alchemy, is again Wisdon, the (soul) in her glorified form...[22]

Thus, Wisdom, which the reader must recall also symbolized God "in his feminine aspect," also appears as the lily, and the rose, two flowers not without their own alchemical meaning, for the lily "symbolized the arcane substance: and was "specifically a synonym for the white, feminine substance,"[23] whereas the red rose symbolized the masculine *differentiation* of that same substance. Thus, the alchemosexual imagery of the *Aurora* the bride becomes the producer of "the white (lily) and red (wine), a hermpahroditic being who unites the opposites in herself. Not only are they contained in her, she is actually the medium of their conjunction, as the next passages shows: 'I am the bed of my beloved...'."[24]

Then, came the most alchemical passage of them all, and it is best to repeat it before citing Von Franz's commentary:

wherefore have all the philosophers commended me and sowed in me their gold and silver and incombustible grain. and unless that grain falling into me die, itself shall remain alone, but if it die it bringeth forth threefold fruit: for the first it shall bring forth shall be good because it was sown in good earth, namely of pearls; the second likewise good because it was sown in better earth, namely of leaves (silver); the third shall bring forth a thousand-fold because it was sown in the best earth, namely of gold....[25]

Von Franz comments at length on the alchemical symbolism of the explicit imagery of this passage:

This motif, too, is taken over from the previous parable, where Hermes commanded his son to sow gold in the earth of the Promised Land. The "earth of leaves"...is the "white foliate earth"...or silver-earth. In Senior, *granum* (grain) is sometimes the (alchemical) tincture, sometimes gold, *and sometimes the soul.* The "Rosarium," commenting on this passage, explains the grain as the "grain of the body" and the earth as the prima materia, which absorbs the "fatty vapour" (Mercurius). In Aenigma VII of "Allegoriae super librum Turbae" the "single grain of burgeoning seed" must be joined in mystic marriage to the "primordial vapour of the earth." The primordial earth-vapour, the fatty vapour, and Mercurius are, accordingly, all synonyms for the Promised Land and show that an airy, sublimated earth is meant. This consists of three substances, pearls, silver, and gold; we find the same classification in Senior...

(And, it is to be noted, in the *Aurora*!)

...we find the same classification in Senior, from whom it was taken over. This mystic earth is therefore a kind of lower Trinity....

The essential thing about this lower Trinity is that it is described as an earth, i.e., as a psychic reality which has to do with the nature of matter. Matter thus acquires an importance of its own and is even raised to divine rank - in complete reversal of the medieval scholastic view, according to which, matter, unless it is given form, has only potential reality. The text is, in fact, proclaiming a glorification of the feminine principle, of the body and matter. *From this we can see what a shattering breakthrough of unconscious contents was needed before a man of the Middle Ages* could hazard such a statement.[26]

Nor is this all.

Let us recall the following passage from the Seventh Parable, cited in the above lengthy quotation:

> I am the law in the priest and the word in the prophet and counsel in the wise. I will kill and I will make to live and there is none that can deliver out of my hand. I stretch forth my mouth to my beloved and he presseth his to me; he and I are one; who shall separate us from love? None and no man, for our love is as strong as death.[27]

Here, again, Von Franz comes to the heart of the matter, and to the heart of why, if Aquinas was indeed the author of this extraordinarily ecstatic treatise, his association with it had to be denied or at least not spoken of too broadly by the ecclesiastical authorities:

> The identity of the bride with God is beyond all doubt, for the words she speaks are God's own (Deut. 32:39: "I kill, and I make alive; I wound, and I heal; neither is there any that can deliver out of my hand"). She *is* God, or his feminine "correspondence" in matter. She is God, but as a loving woman embracing the man in order to draw him into God's own antithetical nature and at the same time into his all-enveloping wholeness. This experience, as the text says, transcends even death.
>
> This (mystical union) is compared with the usual medieval texts, something new and completely different, *because ordinarily it is the human soul as a feminine being that unites with Christ or God. Man or his anima is the bride. Here, on the contrary, God is the bride, and man or the self the bridegroom.*
>
> *This singular **exchange of roles** must be understood in the first place as a compensatory phenomenon: the masculine, spiritual God-image has turned into its opposite, into a figure that unites God's self-reflection"* - Sophia or Wisdom - *with matter and nature.* What is manifest in this figure is an aspect of God which is striving to become conscious of itself - as though the human psyche and matter were the chosen place for God's self-realization. *The son-lover of this figure, however, is a human being in a glorified end-state,* who has passed through death. In contrast to Wisdom, *he* has cast off the darkness from himself....
>
> ...(A) pagan *joie de vivre* breaks through in words that border on the heretical. At the same time the text conveys a feeling of inner liberation, as if the prison of conventional religious ideas and human

narrow-mindedness had finally burst open, and the author had left his previous mental world behind him like an empty shell.[28]

Clearly, the author of the *Aurora* has passed beyond even the *idea* of the androgyny of God or of Man, to the ecstatic alchemosexual practice or *experience* of it, a crucial point to which we shall return in a moment.

And what of the idea of the bed, of the vessel, or bowl, that is so often associated with the feminine in alchemical works?

> The vessel had to be round because it was an image of the cosmos and of the heavenly spheres, as well as of the human head as the seat of the (rational soul). In the *Corpus Hermeticum* the cosmos is described as a vessel or sphere, and this sphere was also the Nous, which moved like a head; everything connected with this head was immortal.... The *krater* (mixing bowl) also has a hylic significance: in the *Corpus Hermeticum*, matter (and in Plutarch, time) is called the vessel of genesis and decay, and in Neoplatonism the cosmos is a hollow or cave. Plato, and later, certain Oprhic sects held that the world-creator mixed the cosmos in a huge *krater*, and Zosimus in his vision saw the elements being transformed in a "bowl shaped" altar which embraced the entire cosmos. Such is the context of the bride's designation as *crater tornalis*.[29]

But why emphasize this idea of the *receptacle* as the cosmic mixing bowl?

a. The Twofold Movement

The bed - the image of the receptacle - results because of the role-reversal of God and Man in the Seventh Parable of the *Aurora*, and the "two-fold movement" that results therefrom. Let us recall that both God and Man are spoken of in the *Aurora* as *both* being the bride, and the bridegroom. In other words, *both God and man in the **Aurora** appear now as masculine, now as feminine, androgynies.* But they do not do so haphazardly. Rather, God appears as a masculine-androgyne when viewed in His role as creator, and Man as a feminine-androgyne in the same context. However, *in the eschatological context of the resurrection and end time, these roles are reversed*: God becomes the feminine passive androgyne, and Man the active masculine one. The blatantly alchemosexual imagery here forms the key to deciphering what is meant, for in alchemical terms, the masculine is always the active principle, and the feminine the passive one, and in both contexts - creation or end time - the activity suggested by the *Aurora* is one of intimate communal love between androgynies, of a restoration of the knowledge and power of perception that once went with that state "wherein is hidden all the science," to quote the end of the *Aurora*.[30] With this, we are once again in the presence of the ages-old idea that androgyny somehow resulted in a higher state of knowledge, exactly as we saw in the first chapter with the Mayan *Popol Vuh* and Plato's *Symposium*.

Von Franz puts this end-time eschatological vision, the agenda hidden in the vision, in the following terms:

The bridegroom...

who, let it be recalled, is Man,

...summons his beloved...

who, let it be recalled, is *God*,

> summons his beloved to go out into the country and celebrate a feast of joy with men, "for the night is past and the day is at hand.".... it brings an entirely new state of consciousness. In this new light the lovers enjoy their bliss.
>
> "Let us therefore enjoy them and use the good things speedily as in youth, let us fill ourselves with costly wine and ointments, and let no flower pass by us save we crown ourselves therewith, first with lilies, then with roses, before they be withered. Let no meadow escape our riot. Let none of us go without his part in our luxury..."
>
> This passage has no ecclesiastical parallel, for the word which the author puts into the mouth of the bridegroom are those spoken by the "ungodly"...in Wisdom 2:5ff. Either he suffered a lapse of memory *or else he was consciously alluding to a non-Christian mystery.* At any rate it is a breakthrough of the classical or pagan feeling for nature, or - to speak with Hercalitus - a "phallic hymn" sung in honour of Dionysius and Hades.[31]

b. The Implications

The implications of all of this are rather obvious, not the least of which, we are bold to suggest, is that if the *Aurora Consurgens* is indeed the commentary on the Song of Solomon that two medieval texts says it is, then this blatantly alchemosexual vision would perfectly and fully rationalize why the *Doctor Angelicus* of the medieval church put down his pen and never wrote another word on Christian theology, choosing only to spend his final days expounding on an altogether different, alchemical transformation of Man and God behind a riot of biblical and alchemical symbols. If indeed Aquinas was the author of the *Aurora*, it is small wonder that so many dispute the authorship to this day, for it would mean that he, like a much later author in the nineteenth century, writing on the same mysteries, had perceived that all science and art were "at once surface and symbol," and that he had gone "beneath the surface" and "read the symbol,"[32] and understood the ancient metaphor behind the obscuring veil of religion.

There is one final contextual piece of the puzzle that also suggests that Aquinas could indeed be the author, and it is a piece that is often associated with his name: the alchemical doctrine of transubstantiation itself.

C. TRANSUBSTANTIATION AND THE TOPOLOGICAL METAPHOR

We believe there is a profoundly alchemical, analogical clue here, and that, once again, it is capable of being captured with the formal explicitness of a topological notation, for what transubstantiation is, is in effect, a *mapping* function from one context to another, but with a peculiarity, in that in this case - if one follows the "annihilation" model of transubstantiation - it is the *topological interior*, what the scholastics called the "substance" of bread and wine that is annihilated, while the *surface*, the "accidents" of bread and wine, that remain the same from one context to another. Let it also be recalled, in this context, that both Thomas Aquinas, and his mentor Albertus Magnus, wrote on alchemy.

If the transubstantiation doctrine were to be notated in terms of a topological recontextualizing mapping function, it might look like this:

$$\{_{x}\text{...}S_{x}\text{...}_x \to _{y}\text{...}S_x - S_x^{\,\circ} + \partial S_x + S_y^{\,\circ}\text{...}_y\}$$

where the resultant on the right side of the arrow reduces to

$$= {}_{y}\text{...}\partial S_x + S_y^{\,\circ}\text{...}_y,$$

where S_x is the total object (substance of bread and wine plus their accidents), $S_x^{\,\circ}$ is the "interior" or "substance" of bread and wine, ∂S_x the surface, or "accidents" of bread and wine, and $S_y^{\,\circ}$ the "interior" or "substance" of body and blood, and where ••• to either side of an expression means "in the context of," in this case, context "x", the secular context, and context "y", the ritualistic or liturgical, and ° represents the mapping or transmutation itself. We have chosen to notate the function in these linguistic contextual terms because in the Latin doctrine of transubstantiation, the transmutation is effected linguistically, by the utterance of the "magical" formula, "*hic est corpus meum*" or "this is my body, this is my blood," and so on. Thus, the result on the right, one surface possessing the interior of another substance entirely, expresses the oxymoronic, alchemical blending that results from the doctrine, which, be it noted, is revealed as a mapping function.

Of course, the problem is revealed by the peculiar notation of the doctrine adopted here, for topologically, it is impossible to preserve the surface of the region in the resultant of the mapping without preserving or implying the interior, or substance, as well, which, in theological terms, was pointed out during the Middle Ages and the Protestant Reformation.

But for the reader who has carefully followed our remarks in chapter two, and in our previous book *The Grid of the Gods*,[33] the problem with the doctrine, especially when notated in this fashion as a context-specific mapping function, will be immediately revealed, for instead of resulting in a *common surface of two distinct things or regions*, the result is a technical chimerical impossibility: a region or interior(a "substance" in scholastic terms) of one thing, with the surface(or "accidents" in scholastic terms) of another thing. It is the very *opposite* or reversal of the alchemical symbol of androgyny, a kind of false alchemy in the guise of an ecclesiastical doctrine. It is, as we observed in chapter two, a form of nihilism.

However, the author of the *Aurora* was not the only one within western culture wrestling with the alchemical symbols of androgyny and its implications. Oddly enough, it was a major theme of nineteenth century men of letters and science, as we will see in the next chapter, before returning to a consideration of the tradition of secret societies, and little-known religious traditions of androgyny within Christianity and Judaism in chapter ten.

Endnotes

1   Processus Canonizationis S. Thomae Aquinatis Neapoli", Pruemmer, *Fontes*, p. 377, cited in *Aurora Consurgens: A Document Attributed to Thomas Aquinas on the Problem of Opposites in Alchemy*, trans. R.F.C. Hull and A.S.B. Glover, Ed. and Commentary by Marie-Louise Von Franz (Toronto: Inner City Books, 2000), p. 425.

2   Dante Alighieri, *The Divine Comedy: Paradiso*, trans. John Ciardi (New American Library, 2003), p. 631.

3   Marie-Louise Von Franz, "Was Thomas Aquinas the Author?" *Aurora Consurgens*, 407-431, p. 425, citing Angelo Walz, *St. Thomas Aquinas: A Biographical Study*, trans. Sebastian Bullough (Westminster, maryland, 1951), pp. 155, 160, emphasis added by the authors.

4   *Aurora Consurgens*, p. 426.

5   Ibid., p. 427.

6   Ibid., p. 4, from the introduction by the translators, R.F.C. Hull and A.S.B. Glover.

7   Ibid.

8   Ibid., p. 5.

9   Ibid.
    Marie-Louise Von Franz, "Was Thomas Aquinas the Author?" *Aurora Consurgens*, p.416.

10 *Aurora Consurgens*, pp. 7-21.

11 Ibid., p. 22.

12 Ibid.

13 Marie-Louise Von Franz, "Was Thomas Aquinas the Author?" *Aurora Consurgens*, p. 408.

14 Ibid., emphasis added.

15 Ibid.

16 Ibid, p. 427.

17 Marie-Louise Von Franz, Commentary, *Aurora Consurgens*, p. 362.

18 "Seventh Parable, "Of the Confabulation of the Lover with the Beloved," *Aurora Consurgens*, p. 133, citing Joel, 2:12: "now therefore saith the Lord: be converted to me with all your heart...".

19 Marie-Louise Von France, "Commentary," *Aurora Consurgens*, p. 363.

20 Ibid., p. 364.

21 "Seventh Parable, "Of the Confabulation of the Lover with the Beloved," *Aurora Consurgens*, pp. 137-149, emphasis added.

22 Marie-Louise Von Franz, Commentary, *Aurora Consurgens*, p. 370.

23 Ibid., p. 376.

24 Ibid., p. 378.

25 *Aurora Consurgens*, pp. 141, 143.

26 Marie-Louise Von Franz, Commentary, *Aurora Consurgens*, pp. 384-385, emphasis added.

27 "Of the Confabulation of the Lover with the Beloved," *Aurora Consurgens*, pp. 143, 145.

28 Marie-Louise Von Franz, Commentary, *Aurora Consurgens*, pp. 387-389, emphasis added.

29 Ibid., pp. 389-390.

30 *Aurora Consurgens*, p. 149.

31 Marie-Louise Von Franz, Commentary, *Aurora Consurgens*, pp. 391-392, emphasis added.

32 Oscar Wilde, "Preface," *The Picture of Dorian Gray* (Barnes and Noble, 2993), p. 2.

33 See Joseph P. Farrell and Scott D. de Hart, *The Grid of the Gods: The Aftermath of the Cosmic War and the Physics of the Pyramid People* (Adventures Unlimited Press, 2011).

FRANKENSTEIN'S ALCHEMICAL FICTION:
SHELLEY UNBOUND AND THE PICTURE OF OSCAR WILDE,
AN INTRODUCTION TO CHAPTER 9.

As we introduce the following chapter, a question naturally arises, why include an analysis of two 19th century works of fiction? The purpose of *Tranhumanism: A Grimoire* is to examine important and neglected works where a hidden agenda was passed from ancients to moderns, from fraternity members to priests and poets; to expose an insidious goal established upon the premise that only through the annihilation of one substance is it possible to recreate a new higher substance, specifically the individual and generally the human race.

This alchemical agenda lurks in the poet's pen no less than in the chemist's cauldron. While it might seem an overreaching stretch of the imagination, the brief consideration of 19th century *fiction* is a critical and overlooked area where transhumanist ideals were hidden in plain sight by men who clearly saw into the future while having a firm hold on the alchemical agenda of the past. Fiction it may be, but the message passed carefully through the author's pen betrays a higher purpose. Both Percy Shelley and Oscar Wilde were poets first and novelists secondarily. Both men were steeped in the belief that "the great poet is always a seer,"[2] and that the foundation of civilization is first realized through the words of a poet, "in the infancy of society," writes Percy Shelley, "every author is necessarily a poet, because language itself is poetry; and to be a poet is to apprehend the true and the beautiful, in a word, the good which exists..."[3] The alchemical agenda takes more than one form of expression and this survey of two classics merely indicates the tip of a literary iceberg; the massive mountain beneath the surface is reserved for a future work dedicated to that single purpose.

ENDNOTES

1 Samuel Taylor Coleridge, *Specimens of Table Talk S.T. Coleridge,* 1835.

2 Oscar Wilde, :The Critic as Artist ," *The Complete Works of Oscar Wilde* (Collins: 2003), p.

3 Percy Bysshe Shelley, *A Defense of Poetry* (), p.

♣ Nine ♣

MONSTERS, PICTURES, AND PITS:
SHELLEY, WILDE, DANTE, AND THE LITERARY REVERSAL OF THE TOWER OF BABEL MOMENT

∴

It was the secrets of heaven and earth that I desired to learn; and whether it was the outward substance of things or the inner spirit of nature and the mysterious soul of man that occupied me, still my inquiries were directed to the metaphysical, or in its highest sense, the physical secrets of the world.
—Mary Wollstonecraft Shelley/Percy Bysshe Shelley, Frankenstein.[1]

A. MONSTERS AND MYTHS

JAMES WHALE, "The Queen of Hollywood", best known as the director of the classic horror film *Frankenstein*, portrayed Shelley's monster as a lumbering, mentally deficient, unintelligible beast. Boris Karloff's famous role as the monster established such an iconic image that few today would be able to conjure up an impression of the monster other than Karloff in platform shoes with bolts in his neck. The cinematic magic created by Whale and Karloff have entertained millions of viewers and can hardly be dismissed. Frankenstein's monster is a Hollywood creation in itself and has left a cultural legacy in celluloid. Regrettably, the 1931 classic horror film bears little resemblance to the novel, its underlying themes, or the shocking implications contained in the early 19th century epistolary novel. Yet, as much as Hollywood has done to imprint the popular impression of the man and his monster, Whale's *Frankenstein* is only one of several potential obstacles effectively misdirecting or obstructing some alarming themes hidden in plain sight within the novel.

The long standing tradition associated with *Frankenstein's* authorship, perhaps more than anything else, has obstructed key themes lurking in the novel. Surprisingly enough, the question of authorship is not a new controversy, as from the moment the novel was written the question of authorship was in question. Why? Initially *Frankenstein* was published anonymously. Newspapers and literary scholars, no less than Sir Walter Scott, attributed the anonymous novel to Percy Bysshe Shelley rather than Mary Wollstonecraft Godwin (Shelley). The fact is that this young man, a pacifist and social reformer, poet and atheist, had a penchant for writing anonymously or with a pen name. It would hardly have come as a surprise that Shelley would contact a publisher with the *Frankenstein* manuscript and request that the author remain anonymous. Only *after* speculation surfaced that Shelley was the author did he begin to set forward the idea that Mary was the actual author. Given the substance and amount of published material from Percy Bysshe Shelley prior to *Frankenstein* (published in 1818) it took some convincing on the part of the public to believe that Mary, at the tender age of 18, was capable of writing a first book with such skill and touching on such scientific and speculative themes.

Today, it is common place in academia to accept Shelley's word that he was not the author; case closed. However, the question of authorship is worthy of a second look given the life and publishing pattern of Shelley; in fact it is certain to yield rewards regardless of where one finally settles on this matter so long as it is *seriously* reconsidered. It is our view, based on the weight of certain comparative textual and particularly biographical evidence, that Percy Bysshe Shelley is not only responsible for the inspiring the idea of *Frankenstein* but also was the primary author behind the scene, dictating or writing in his own hand most of the original manuscript, while allowing Mary Wollstonecraft Godwin to launch her career as a writer through his ideas and words. Shelley's willingness, even desire to hide his involvement added to the confusion over authorship. Recent scholarship by Charles Robinson has credited Shelley with approximately 4,000 words in the Frankenstein Notebooks (manuscript) but this is significantly misleading as it assumes that *handwriting is the evidence* of authorship; an argument which quickly breaks down when an author such as Shelley often dictated his work and/or had someone else (often Mary) transcribe! Furthermore, Shelley's literary fingerprints, i.e. thoughts, are more evident than any blot of ink leaked from the pen of Mary Wollstonecraft Godwin in the Frankenstein Notebooks. A thorough analysis requires far more than handwriting analysis as the work behind *Frankenstein* is as mysterious and private as the creature's maker Victor Frankenstein.

Additionally, creating an uncertainty of authorship was Shelley's modus operandi throughout his publishing life, a calculated, and intentional misdirection, when dealing with controversial subjects. Nonetheless, Mary Wollstonecraft Shelley's credit for being the author of *Frankenstein* has not only remained in place, as Shelley most likely expected, but it has become an academic taboo to question Mary's authorship. We challenge and intend to raise doubt over the accepted approach to a canonical Mary Shelley authorship and will propose several credible reasons as to why Shelley might have thought it necessary or useful to conspire and hide his name as the author of *Frankenstein*.[2] There is more that meets the eye in *Frankenstein* than most readers have considered. Our goal is to open a window of further insight by raising challenges, questions, and providing some probing solutions.

Frankenstein is a masterfully crafted novel whose message is more relevant today than when it was first published in 1818. With advances in science, particularly in genetics, Shelley's novel takes on a contemporary and frighteningly greater relevance. Though fictional, *Frankenstein* must be re-approached with a clear understanding of Shelley's scholarly and fascinating treatment of alchemy and theoretical science which he covertly weaved with the threads of his own social ideals and personal experiences. *Frankenstein*, an early 19th century epistolary masterpiece of *fiction* may be looked upon today as a 21st century work of *non*-fiction and in it Shelley has cleverly pointed out the possibility for a new civilization… the final civilization! In the words of the tormented creator of his "monster", Victor Frankenstein exclaims,

> I shuddered to think that future ages might curse me as their pest, whose selfishness had not hesitated to buy its own peace at the price perhaps of the existence of the whole human race …a race of devils would be propagated upon the earth from whose form and mind man shrunk with horror.[3]

Indeed, Shelley was fully aware of the potential for evil as well as progress rising to the surface as science was gaining an authoritative voice against religion. Overturning social structures, particularly the Christian religion, might well have been his hope weaved into the pages of *Frankenstein*. Victor Frankenstein represents a Shelleyan idealism for human advancement, an advancement quite impossible if intellectually enslaved to Christian theology. Radical independence, even if ethically unorthodox, was to Shelley a means towards unity or universal oneness. Such seemingly contradictory or opposing forces drove the creative genius of Shelley's work and likewise created a push- pull effect in his relationships with men and women. To understand

Frankenstein is to enter a complex spider's web with delicate strands of ethics and philosophy, nearly invisible to the naked eye, attaching one idea to the next, all leading back to the creator of the web, its author.

In considering the evidence for Shelley's agenda, one that is clearly alchemistic, Neo-Platonistic, and scientific, all converging within *Frankenstein*, we are placed in the unenviable position of having to reduce the argument for his predominant authorship to a few paragraphs lest the chapter's primary aim is lost for the purpose of proving an important but secondary issue to the present book. Why has Shelley's role in authorship been ignored by previous studies? Simply, the vast majority of modern studies of *Frankenstein* have been so committed to the accepted idea of Mary Wollstonecraft Shelley's entire authorship that a practical blind spot has set in where literary analysis is concerned. Among the very first lessons that an English student is taught to consider is authorship and context before attempting to analyze literature. *Frankenstein* presents the greater challenge that its author attempted to be anonymous, meaning that some degree of deception was being perpetrated from the start. The announcement of Mary as its author proclaims "mystery solved" ... but are there unresolved issues still and do they matter? Yes and yes. So now we ask, what evidence is given for authorship?

If one accepts the standard view that Mary Wollstonecraft Shelley was the author of *Frankenstein*, an answer regarding authorship and inspiration is conveniently provided. According to Mary, in 1816, on a stormy summer day in Switzerland where she, Percy, and Lord Byron (and other companions) were staying, a contest of sorts was suggested by Lord Byron, that each of those present should write a ghost story. Byron's contest took hold of her and from that moment onward she "busied herself with thinking of a story." A more detailed account of Mary's inspiration may be found in her Introduction to the **1831** highly revised Edition of *Frankenstein*:

> I busied myself to think of a story, - a story to rival those which had excited us to this task. One which would speak to the mysterious fears of our nature, and awaken thrilling horror – one to make the reader dread to look round, to curdle the blood, and quicken the beatings of the heart.[4]

Assuming for the moment, Mary Shelley's predominant authorship and her account of its origin (rather than a guided transcriber and minor contributor to the novel), *Frankenstein* is nothing more than a ghost story. Its intention was to be a contest winning horror story; an awakening, thrilling, even blood-curdling, heart-stopping story. As such, one would not expect to find the author burying indiscreet or dangerous secrets; nor would it be a mani-festo for change by alchemical means; nor would it reflect such controversial figures or topics such as the Illuminati, hermaphroditicism, and Paracelsus, all of which are found in *Frankenstein* and repeatedly in the pages of Percy Bysshe Shelley's life, poetry and prose, i.e. *Prometheus Unbound; The Witch of Atlas; The Sensitive Plant; Epipsychidion*. As a point of fact, let us be perfectly clear and allow Mary to speak for herself as she argues that the story was *not even suggested* by anything in relationship to her husband Percy. Mary declares, "I certainly did not owe the suggestion of one incident, nor scarcely on one train of feeling to my husband."[5] This is undoubtedly one of the most damning falsehoods ever penned, unless our assumption is near the truth, that Mary's authorship was an idea that Shelley himself concocted and she perpetuated with his explicit approval. Under Mary's hand then, *Frankenstein* is but a flight of fancy, a spark of her creative imagination, it is by her own account nothing more than her own ghost story.

Is it *possible* that this young girl, at the age of 18, wrote *Frankenstein*? In a word, yes. Is she *the most likely* author given the subject matter and circumstanc-es? In a word, no. Briefly consider a few relevant facts: Mary Wollstonecraft Shelley had significantly less academic background than Shelley, often receiv-ing her education on the spot from Shelley; she had far less experience and acquaintance with Greek and Latin philosophy than Shelley; showed consid-erably less interest in Illuminati agendas, the alchemists, and theology than Shelley; additionally, Mary was not formally schooled in science and anatomy as was Shelley, both before and after Shelley's brief University of Oxford expe-rience. It is highly unlikely, given a comparison of the two potential authors, that Mary Wollstonecraft Shelley could *or would* write a novel so utterly de-pendent upon a mastery of the subjects as are contained in *Frankenstein*. Mary was generally lacking in those foundational ideas or philosophies which were abundant in Shelley's educational, professional, and personal life, particularly as they relate to the themes contained in *Frankenstein*. If, as we argue, there is a deeper meaning to *Frankenstein*, Shelley alone could conceive of the real possibilities inherent in man creating or recreating human-*like* life through a proper application of esoteric teaching, alchemy, and science.

Mary Shelley's inspiration, she professes, did not come from her high education, years of personal research, and the experiences of wrestling within herself over such matters of alchemy, the creation of life without God's in-tervention, and even reforming social and gender roles in society. Her in-spiration came, as she describes, from eavesdropping on the deep discus-sions that *others* were having on such matters, and one may correctly assume, discussions by *informed* men who understood the implications! Again, Mary

Wollstonecraft Shelley describes her supposed inspiration within the 1831 edition of *Frankenstein;* the edition that was published with her name as author after the death of Percy Bysshe Shelley*:*

> Many and long were the conversations between Lord Byron and Shelley, to which I was a devout but nearly silent listener. During one of these, various philosophical doctrines were discussed, and among others the nature of the principle of life, and whether there was any probability of its ever being discovered and communicated. They talked of the experiments of Dr. [Erasums] Darwin, (I speak not of what the Doctor really did, or said that he did, but, as more to my purpose, of what was then spoken of as having been done by him).[6]

As should be obvious, Mary has just contradicted herself regarding *any* influence from Shelley. Furthermore, note that Mary, driven by a whimsical wager, "busying herself" thinking of a story, and then overhearing a philosophical discussion between Byron and Shelley, Mary supposedly crafted her novel detailing the exploits of Robert Walton, the risk taking sea farer and explorer; Victor Frankenstein, a student of theoretical science who was educated at Ingolstadt; and of course the alchemically inspired creature born without the aid of a woman. We have already touched upon such alchemistic themes arising in previous chapters, homunculi of the past and modern experiments which exclude the participation of a woman's contribution to birth. Such topics demand a detailed and well researched book unto itself rather than a few paragraphs with a summary argument in the present book to *fully* explore the significance of the problem of authorship, themes, and the secrets behind Shelley and his literary output (including *Frankenstein*) but suffice it to say the matter is hardly a closed case as it is suggested by the vast majority of studies on *Frankenstein.*[7]

We recognize the over simplification of such a chain of events leading to the origin of one of the greatest and often underestimated novels written in the English language, however at face value it is a succinct analysis from which to view the novel *if* written by Mary Wollstonecraft Shelley. We unapologetically depart from *popular* academic opinion on this point of authorship (though our position is on an academic footing) as well as departing from the resulting literary criticism which is in keeping with such an origin for *Frankenstein.* We do not argue that there is a lack of value in academic contributions by previous studies which have adopted the entire Mary Shelley predominant authorship, however it must be granted that ***if** Frankenstein* was the product of an author other than Mary, then there is a gaping hole in the literary analysis which has been produced by scholars for nearly 200 years. Such a hole seems not only apparent but glaring.

Challenging the status quo or accepted academic opinion within any discipline opens one up to criticism or dismissal, however every major academic breakthrough begins with a new thought, a *dangerous* and challenging idea. In fact, it would be an act of ingratitude to write about Shelley and then not challenge accepted ideas, which in the very act of challenging produce potentially valuable ideas. The intention of this chapter *is* to challenge previous ideas, to purposefully reexamine concepts that have been accepted in popular opinion through the media of Hollywood as well as to cross the proverbial line in the exclusive academic sandbox with regard to *Frankenstein* analysis.

We do face these previously mentioned challenges and the reader must overcome them in order to move forward and see *Frankenstein* for what it actually contains. First, there is James Whale, the director of the 1931 *Frankenstein* horror classic. He chose this particular story out of all that were offered to him at Universal for personal reasons (another book for another time); Whale interpreted the story in his own magnificent fashion, created the iconic monster, added homunculi in the sequel *The Bride of Frankenstein*, and left his fingerprints on the public perception of *Frankenstein*. With gratitude we raise our box of popcorn to Mr. Whale and Mr. Karloff, but ask the reader to disregard this image! Additionally, with due appreciation to Mrs. Mary Wollstonecraft Shelley for her important work as a transcriber and perhaps also as a minor contributor to the 1818 edition of *Frankenstein*, we ask the reader to set aside his or her certainty of Mary's authorship and consider the possibility that another more likely and informed hidden author was behind the book. The reader is asked to consider that the intended veil over Shelley's predominant authorship was designed by Shelley himself, that he initially used an *anonymous* authorship to cast doubts away from revealing himself or any agenda attributed to him, and then later attributed full authorship to Mary Wollstonecraft Shelley as a means to launch her career as well as a means to disguise any part of his hand in the book, other than slight editorial suggestions. The authors and readers are now faced with breaking two time honored molds as related to the novel, the *in*credible Mary Shelley authorship of a book largely created as a "ghost story," and the James Whale iconic, dim-witted, clumsy mute giant who is driven to perish in a burning Hollywood-set windmill. The true monster, at least as society was concerned, was Shelley himself. Let us be certain, the creature did not consider himself a monster, though undoubtedly accepted his singularity in society.

B. The Monster Within the Man:
A Promethean Literary Alchemistic Agenda

Percy Bysshe Shelley, the man and soul behind *Frankenstein*, was largely an outcast as a child, a virtual stranger to his own gender while in public school, a man of tremendous emotion, conviction, and passion. He was thought of as immoral, atheistic, and delusional. He was an advocate for equality, libertinism, reform, and deism/atheism.[8] He was a scholar, a poet, and an enthralled metaphysical visionary who risked his life, livelihood, and reputation rather than compromise his worldview. John Addington Symonds writes of Shelley that,

> His genius by a stretch of fancy might be compared to one of those double stars which dart blue and red rays of light: for it was governed by two luminaries, poetry and metaphysics.[9]

Shelley's childhood foreshadows the very story that would become a horror classic. Shelley's sister describes him, even while still a young boy, as entertaining them with stories of "an old and grey bearded *alchemist*," one of his favorite subjects and imaginary characters. Another recollection from family life in the Shelley household describes "some of the games he invented to please his sisters were grotesque, and some both perilous and terrifying. [one sister remembering] We dressed ourselves in strange costumes to personate *spirits or fiends*..."[10]

Shelley was equally remembered for his kind and gentle spirit, even effeminate demeanor and appearance which did *not* endear him to his own peers while in school, making him an outcast and subject for reviling – he was an icon of androgyny. He was not like the other boys and his interests did not draw friends to him, rather he found himself more comfortable alone or with a singular friend. He did not participate in sports and was more likely to be independently engaged in writing verses, reading, or scribbling sketches of nature. He had neither respect nor desire to keep company with those who used violence or superstitious threats to control others; for establishments which arbitrarily – or in the name of God – set out the rules of behavior in relationships; and therefore Shelley determined at the earliest age to be independent and learn from those writers in the past whose lives and ideals were most like his own. This meant that Shelley would stand apart, be ridiculed, be judged, risk banishment, and suffer loneliness, but he would, even at a great cost to himself and his reputation, raise a voice against what he saw as injustice in his day.

In the Prelude to *Laon and Cythna*, Shelley is likely reflecting on his own childhood when he writes,

… so without shame I spake: - "I will be wise,
and just, and free, and mild, if in me lies
Such power, for I grow weary to behold
The selfish and the strong still tyrannize
Without reproach or check." I then controlled
My tears, my heart grew calm, and I was meek and bold.

And from that hour did I with earnest thought
Heap knowledge from forbidden mines of lore,
Yet nothing that my tyrants knew or taught
I cared to learn, *but from that secret store*
Wrought linked armour for my soul, before
It might walk forth to war among mankind.
Thus power and hope were strengthened more and more
Within me, till there came upon my mind
A sense of loneliness, a thirst with which I pined.

Shelley's determination, even in his youth, to heap knowledge from the forbidden intellectual and philosophical storehouses, to seek light, even hidden (Hermetic) wisdom, and turn it into the inspiration for his life's philosophy, surely must have also laid the foundation for writing a novel whose protagonist was well studied in alchemy, theology, and science. In James Bieri's authoritative biography on Shelley, Bieri notes that while Shelley was at Eton, he brought with him an electrical machine (one he used to experiment on friends and family); bought chemical apparatus; obtained books on magic and witchcraft; drank from a skull; and was tutored by Dr. Lind, considered the modern day Paracelsus. Shelley's favorite topics for research were chemistry, magic, alchemy, and the writings of Paracelsus.[11]

Victor Frankenstein, recounting his childhood, tells his chronicler Walton,

When I was eleven years old we all went to a party of pleasure to the baths near Thonon. The inclemency of the weather obliged us to remain a day confined to the inn. In this house I chanced to find a volume of the works of Cornelius Agrippa … A new light dawned upon my mind. I continued to read with the greatest avidity. When I returned home, my first care was to procure the whole works of this author and afterwards those of Paracelsus and Albertus Magnus. I read and studied the wild fancies of these authors with delight; they appeared to me treasures known to few besides myself and although I often wished to communicate these secret stories of knowledge to my father, yet his definite censure of my favorite Agrippa always withheld me. I disclosed my secret to Elizabeth, therefore, under a strict promise of secrecy; but she did not interest herself in the subject, and I was left by her to pursue my studies alone ... I entered with the greatest diligence into the search of the philosopher's stone and the elixir of life.[12]

Take careful note of the names Agrippa, Paracelsus, Magnus! Each man a revolutionary in his field, playing with the ingredients and mysteries of life, perhaps even creating or manipulating the forces of nature to create life. Victor Frankenstein, the monster's creator, arises as a virtual autobiographical image of Percy Bysshe Shelley himself. The poetically autobiographical words "a sense of loneliness" and "thirst" consequential to such a life of esoteric study and imagination describe the life of Victor Frankenstein as much as that of Shelley. A price must be paid for a life invested in such controversial studies whose ends often lead to challenging the authority of God.

Symonds describes Shelley's early years in shockingly similar words,

The months which elapsed between Eton and Oxford were an important period in Shelley's life. At this time a boyish liking for his cousin, Harriet Grove, ripened into real attachment ... Shelley and Miss Grove kept up an active correspondence; *but the views he expressed on speculative subjects soon began to alarm her. She consulted her mother and father, and the engagement was broken off.*[13]

So it was that Victor Frankenstein was to live a life of virtual isolation from the one woman in his life on account of his research and experiments pertaining to "speculative matters."

Alas! to me the idea of an immediate union with my cousin was one of horror and dismay. I was bound by a solemn promise which I had not yet fulfilled and dared not break; or, if I did, what manifold miseries might not impend over me and my devoted family! Could I enter into a festival with this deadly weight yet hanging round my neck and bowing me to the ground?[14]

Victor Frankenstein is not the only association that Shelley has with a character in the novel. The creature, the fiend, the cursed monster, the

demon, is no less than Shelley a kind hearted, misunderstood, highly intelligent though lonely being. The creature, pleading with De Lacey, the old and
blind Frenchman in his cottage, exclaims

> I am an unfortunate and deserted creature. I look around and I have no
> relation or friend on earth … I am full of fears … I am an outcast in the
> world forever… I have good dispositions; I love virtue and knowledge;
> my life has been hitherto harmless and in some degree beneficial; but a
> fatal prejudice clouds their eyes; and where they ought to see a feeling
> and kind friend, they behold only a detestable monster.[15]

After being driven by hatred and prejudice into complete isolation, filled with
loneliness and desperation, the creature locates his creator and attempts to
persuade Frankenstein to rescue him from this plight … or to suffer the consequences with him,

> My vices are the children of a forced solitude that I abhor, and my
> virtues will necessarily arise when I shall receive the sympathy of an
> equal. I shall feel the affections of a living being and become linked
> to the chain of existence and events from which I am now excluded.[16]

The creature makes his confession to De Lacey and issues his demands to his
creator. Additionally, the "monster", as viewed from the Maker's point of view,
makes clear that he requires an equal and yet has found none; it was not he who
made himself! Shelley, in this moral dilemma, is revealing his own deepest needs
as well as his indebtedness to Greek ideals regarding love and relationships.[17] *This
point is a critical clue as to an alchemically oriented agenda behind Shelley's novel.*
While it is no secret that Shelley's views regarding marriage were fundamentally
radical, adopting a philosophy of free love, i.e., "open marriage," and that Shelley
even experimented with it in both of his marriages, the real scandal was his deepest struggle personally to make sense of and find his ideal soul mate; the androgynous being most like himself, a concept he derived from Plato. While Shelley
was clearly no stranger to intimacy with the opposite sex, he was equally never
able to find a true complimentary female whose lasting impression (rather than
initial idealistic infatuation) merged fully with his ideal. The woman he sought
as the "other half", it seems was a longing from his own nature. This conflict and
fascination for the ultimate ideal of a person was the Greek androgynous man.
It is not surprising therefore that some of Shelley's most inspiring and lasting
relationships were with men like himself; often men whose sexuality was (by
18th century as well as modern standards) socially unacceptable, deviant, and in

his own times punishable by death for certain behaviors.[18] Clearly Shelley found that such men were the most likely to tolerate him, the androgynous and erratic poet. Shelley, the man, his relationships, and his vision was based on the ancient and it was a glimpse of the Eden he would have on earth, but the world he lived in viewed his ideals as deviant and monstrous. Shelley would strive to give birth and reality to his ideal person and utopia, but each time he made progress it seemed that his efforts were rewarded with short lasting reward. Shelley constantly had women in his life who influenced him, but these relationships were either incestuously driven (as his love for his sister Elizabeth) or were idealistically impossible to carry on for more than brief periods.

Analyzing the complex and interrelated relationships between the author of *Frankenstein* (Shelley), the protagonist (Victor Frankenstein), and the antagonist (the creature/monster), unveils certain revelations about Shelly's own view of friendship, loneliness, virtue, and vice.

Victor Frankenstein speaks with moving affection for his "beloved" friend Henry Clerval, when he confesses "…in Clerval [Henry] I saw the image of my former self." To see ones image in the life of another is a critical component to understanding Shelley's Greek ideal, which will be treated in more detail in the chapter. Frankenstein's feelings for Henry are reciprocated by his friend, who exclaims his own feelings with broken heart at Frankenstein's departure from him,

> Hasten then, my dear friend, to return that I may again feel myself somewhat at home, which I cannot do in your absence.

This relationship snapshot between Victor and Henry is much more than poker playing friends watching Monday night football and knocking down some Budweiser. Male companionship, as described between Victor and Henry is conceived at a higher level of love, a love that Shelley idealized in much of his poetry. With Shelley relationships are self-reflections, mirrors, a potential narcissistic love were it entirely self-centered, but it is far more than that. What is coming into view is Shelley's Greek ideal of an androgynous soul-mate creature seeking its other half.

Victor Frankenstein's relationship to women on the other hand is more reserved, passive; one might even interpret his feelings in relationship with women as more *obligatory* than *necessary*. When contemplating his marriage (arranged and thrust upon him by his father) to his to-be bride Elizabeth, he describes the future union as an idea of "*horror and dismay.*" Frankenstein has been warned that his creature will be with him on his marriage night and while this might account for his thoughts of 'horror and dismay,' yet this argument alone cannot account for Frankenstein's passive emotions towards marriage. When deciding to move forward with the decision of marriage, his reason is clearly passive and impersonal,

> I resolved therefore that if my immediate union with my cousin would conduce either to hers or my father's happiness, my adversary's threats against my life should not retard it a single hour[19].

One might legitimately ask what bride could be persuaded by a proposal such as given by Frankenstein, amounting to, "If it makes *you* happy and knowing it will make my father happy, I guess there is no reason to *not* get married."

Is there a corresponding event in Shelley's life and in his marriages? In a letter to his closest friend, Thomas Jefferson Hogg, Shelley writes of his feelings after marrying Harriet, his first wife,

> I saw the full extent of the calamity which my rash and heartless union with Harriet ... had produced. I felt as if a dead & living body had been linked together in loathsome & horrible communion.[20]

A careful reading of *Frankenstein* unfolds male characters in search of a higher relationship with male friendships while male-female relationships are either passive or non-intimate. The novel launches into this overlooked theme from virtually the opening words of its narrator, the explorer Robert Walton writing to his sister

> But I have one want which I have never yet been able to satisfy; and the absence of the object of which I now feel as a most severe evil. I have no friend, Margaret: When I am glowing with the enthusiasm of success, there will be none to participate in my joy; if I am assailed by disappointment, no one will endeavour to sustain me in dejection... I desire the company of a man who could sympathize with me; whose eyes would reply to mine. You may deem me romantic, my dear sister, but I bitterly feel the want of a friend. I have no one near me, gentle yet courageous, possessed of a cultivated as well as of a capacious mind, whose tastes are like my own, to approve or amend my plans. How would such a friend repair the faults of your poor brother.[21]

Notice once again that a key male character's only relationship is with a relative, a mirror image bordering between incest and narcissism, but Walton's need for her reflection is a correspondence with his own inner demons that have sent him exploring regions of the globe that none others have risked their life to reach.

Not long after writing the previous letter, when Walton finds and saves the wandering Frankenstein, he breaks forth with the words,

> I said in one of my letters, my dear Margaret, that I should find no friend on the wide ocean; yet I have found a man who before his spirit had been broken by misery, I should have been happy to have possessed as the brother of my heart.[22]

Walton seeks a perfectly corresponding male friend and finds him in Victor Frankenstein. Frankenstein in turn has but one dear friend in his life and it is Henry Clerval. The creature seeks out a single equal, the "other half" to live out his life with, to be understood by even if she is as singularly unacceptable to society as he. Initially it is his male creator that the creature seeks such acceptance from, next it is the elderly blind Frenchman in the cottage, and only *after* complete rejection from the men that he justly might expect to embrace him for who he is, does the creature beg for another *like him* in creation, another *creature*, though resolved this time to be female. Again, the image of either narcissism or a fulfillment of the Greek androgynous ideal, the original union of the human person.

There is an undeniable curious attraction towards what is unconventional, even relationally, throughout *Frankenstein*. "Like seeks after like." It is evident that the primary characters have a single aim and it is found only in breaking out of the societal norms, whether as an explorer to lands undiscovered (Walton), or creating life from death (Frankenstein), or uniting the abnormal or "extra-normal" with the same (the monster). In each case, the characters face obstacles in accomplishing their goals and opposition from those they are surrounded by. Walton faces near mutiny from his crew during the expedition to the North; Frankenstein is opposed by his father and professor for his interest in alchemy; the monster is opposed by all of creation for his unnatural place in creation. Each man or monster must break with those who they might have been in harmony with in order to reach their full potential, their highest achievement, in a sense to become perfect or one. The alchemical theme underlying this early 19th century novel is now self-evident; each man's destiny is fulfilled in uniting himself to a greater unifying principle. For Walton it is found in nature and wanting in a friend like himself. For Frankenstein it is found in breaking the mysteries of creation without the aid of god or a woman, and yet his life only finds its moments of fulfillment in the relationship he shares with Clerval. For the monster, it is found in a union *not* with his maker but rather with that which is like him, of one nature with him, relieving him of his knowledge that his existence is a solitary one - unnatural.

C. Alchemy and Frankenstein's Meaning

*"I entered with the greatest diligence into the search of
the philosopher's stone and the elixir of life."*

Although it might at first glance seem off topic in a book concerning Transhumanism with an alchemistic agenda to toss in a section on 19th century literature, it is nonetheless quite necessary and advantageous. Consider the 19th century as if were a god-like marble statue with one foot planted firmly in the past, another foot in the soft soil of the present. Its marble face between the two ages revealing in its cryptic grin how to achieve greatness in the present by a careful understanding and appreciation of the past. Such was the 19th century, a curiously enlightened period in which poets looked to the esoteric past while building an intellectual edifice for their present age. A discerning eye will see that the scientists, psychologists, poets, priests, and revolutionaries of that period were burdened with a mystical calling to accomplish great things as they studied and *read* from the past, particularly the alchemical past! The record imprinted from past writers is a memory of the consciousness of its thinkers, dreamers, its magician-scientists. The stream of consciousness flowing from ancient alchemy past to present runs deeply through the fiction and poetry of the 19th century ... in a way that it never has since that time. Percy Bysshe Shelley was the quintessential product of that period: a stranger; a sojourner as if from ancient times; a cryptic writer, a pacifist with peaceful revolution on his mind; a man whose life would straddle the past with all of the frustration of trying to speak in a language to his contemporaries. It is no wonder he died at the age of 29.

The ancient ideas of alchemy and androgyny run deeply in the thoughts of Shelley. *Frankenstein* must be read as the imagination of Shelley reaching to explain his desire for the perfect form, removed from the necessity of god, and a truly human possibility through the wisdom of modern science matched to the illumination of the alchemists, philosophically webbed by Neo-Platonism. The androgynous person represented for Shelley the union of nature, perfection, oneness and peace. The androgynous ancient person was also a model for the ideal society; the goal of all human relationships. The model as such is revealed time and again in his poetry, particularly in *The Witch of Atlas*. The androgynous witch herself the creation of an alchemical wedding of the Sun and the Moon, a golden cave with secret scrolls and life giving waters surrounds the witch. Her creation calls all of nature to her presence and from her is miraculously born a hermaphrodite. The world in peace, unity, and balance, comes by a sort of alchemistic dream and gives birth to androgyny. In like manner, *Frankenstein* must be seen as an alchemically inspired tale towards a perfect being, a new society, a civilization whose Promethean reach has exceeded god. The experimental creature goes astray but even here Shelley is likely critiquing his fellow visionaries and their failed *Illuminati goals*. Recall that Victor Frankenstein, a student at Ingolstadt (the home of Adam Weishaupt and the Bavarian Illuminati) was set on an irreversible path towards the creature's incarnation while reading the works of Medieval alchemists, Paracelsus, Cornelius Agrippa, and Albertus Magnus. The city, the studies, the alchemical goals for man and society being are apparent immediately.[23]

It has already been shown earlier in the chapter that this experience is virtually identical with the evidence given by Percy Shelley's sister, when young Percy would invent an old grey bearded alchemist as part of his childhood fantasies. Shelley was so well read in Illuminati doctrine and history that *the Code of the Illuminati*, written by legendary author Abbé Augustin Barruel was one of his favorite books to carry and read to others. Erasmus Darwin's *The Temple of Nature*, might even be considered the key influence behind *Frankenstein* if we take the anonymous author's preface at face value. Incidentally, Mary revealed that the Preface was written by Percy when she revised the entire novel and preface in 1831. The significance of parallel occurrences cannot be overemphasized as it was the youthful fascination to find a Philosophers' Stone that led Frankenstein to create life and it was the same which prompted young Shelley to "Heap knowledge from forbidden mines of lore," and it was "from that secret store" that Shelley built the armor for his soul.

Thus, to separate the lives of Frankenstein and Shelley requires more effort than to see the obvious parallels which were autobiographically written between the lines of *Frankenstein*. It was Shelley, in fact, rather than a fictional Victor Frankenstein, who was exploring a means to personally realize the goal of the alchemists. He was also insightful enough to understand that the goal of the alchemists was never gold, but rather *the elevation of man to be one with the organizing principle of all, the divine simplicity or Oneness*!

A deist or atheist such as Shelley (he claimed both at various times in his life) would likely have been *more* driven to find an alchemical transformation *through* knowledge and elevated experience than those who *had accepted* unverifiable non-scientific truth taught by priests and bishops who passed along answers to life's mysteries through the traditions of their religion or god. That "storehouse" or "knowledge from forbidden mines of lore" was to be Shelley's bible, his Beatrix, his guide in search of the holy grail, the final resting place where his ideals of justice and equality for all meet in an eternal Oneness or Simplicity. It is the refrain of the creature in *Frankenstein*, and the societal monster within Shelley, that argues, "I am alone and miserable.."[24]

In such a quest as this it is not surprising that wedded to Shelley's study of Paracelsus, Albertus Magnus, and Cornelius Agrippa, would be none other than Plato.[25] It is in the Greek philosopher Plato that Percy Bysshe Shelley would devote much of his time reading, and later in translating, compelled particularly by his passion to attain the promise of unity, elevation and one-ness through the philosopher's stone. Division and separation violate the internal sense of order and love that motivates Shelley. It is evident in the solitariness that Frankenstein feels after the loss of Clerval. It is evident in the torture that the creature endures every day of his existence, "Man will not associate with me," the creature proclaims, and there is only one solution to the problem – perfect unity, oneness, a joining of common natures together in such a manner that the distinctions end and a oneness arises, "but one as deformed and horrible as myself would not deny herself to me."[26]

This passion for unity, the joining of like nature, a necessity to return or reunite with an inherent Oneness, is instinctively part of the human composi-tion, and is founded in the *eternal* principle of Oneness, perfect divine sim-plicity, a view consistent with the fact that Shelley is an indebted disciple of Plato and the Greek ideal of love. The ideal person is a reality of the past and a potential for the future, with the aid of alchemy and science. Shelley's convic-tion, worked out so carefully in Frankenstein is expounded in a much more straightforward essay, *Manners of the Ancient Greeks Relative to the Subject of Love*, which was combined with his translation of Plato's *The Banquet* (more commonly known as *The Symposium*).

> Let it not be imagined that because the Greeks were deprived of its legitimate object, they were incapable of sentimental love; and that this passion is the mere child of chivalry and the literature of modern times. This object, or its *archetype, forever exists* in the mind, *which selects among those who resemble it, that which most resembles it;* and instinctively fills up the interstices of the imperfect image, in the same manner as the imagination moulds and completes the shapes in clouds, or in the fire, into the resemblances of whatever form, animal, building, etc., happens to be present to it. Man is in his wildest state a social being: a certain degree of civilization and refinement ever produces the want of sympathies still more intimate and complete; and the gratification of the senses is no longer all that is sought in sexual connexion (sic). It soon becomes a very small part of that profound and complicated sentiment, which we call *Love, which is rather the universal thirst for a communion not merely of the senses, but of our whole nature,* intellectual, imaginative and sensitive;

and which, when individualized becomes an *imperious necessity*, only to be satisfied by the complete or partial, actual or supposed fulfillment of its claims.[27]

Notice the human need to find a communion with that which is recognized for its likeness leading to oneness in nature, an embodiment of the archetypal idea of Oneness. For Shelley it does not preclude a sexual element but it is also *higher than something sexual*, it is a perfect love because it is without any sense of division or separation from who or what one is. A relationship that expresses itself only in a sexual manner actually suggests a lower or lesser love, the unity is far deeper. If one wonders how Shelley could be so bold with this essay and so veiled in his novel, the answer is simple; Shelley had no intention that either the essay on the *Manners of the Ancient Greek* or his translation of Plato's *Banquet* should be published at all. As John Lauritsen writes, "such ideas could not be discussed openly in England. Until the middle of the 19th century, males in that benighted country, including adolescent boys, were hanged for having sex with each other." [28] One might still wonder why Shelley was compelled to translate and copy a work he could not risk publishing.

It should come as no surprise that Shelley would not attach his name to *Frankenstein*. Let us speculate a few reasons: He was tossed out of the University of Oxford after publishing (anonymously) a short tract on the *Necessity of Atheism*, thus his name does not lead to *greater* sales but rather a prejudiced condemnation of any writing he produces regardless of merit; *Frankenstein* could be interpreted, if known as Shelley's work, as inspiring an Illuminati anarchist movement and only increase suspicion that Shelley was himself an anarchist; the male characters in *Frankenstein* could be seen as latent homosexuals by the words, actions, and motives which are equally in clear violation of the Christian idea that man was made in the image of God (and not the fancy of alchemists or theoretical scientists). Our suspicion is, as with so many of Shelley's works, that Shelley was not interested in name recognition but he was interested in effecting a reaction. He wrote autobiographical poems and prose, more often than not wanting to hide himself behind the image of his characters and story. Poetry and prose was, for Shelley, the ideal world in which his desires and hopes were lived out. Frankenstein is a novel with Shelley embedded within every page, from desire to frustration. Publishing anonymously was also, at times, a type of voyeurism for the poet to act up in his books through the characters, write himself into his fantasies, disguise his authorship, and enjoy the thrill of others reacting to the story, disgusted, angry, or aroused. We must not forget that Shelley's openness to ménage á trois in the bedroom was only a more graphic representation of his personality than his covert writing. Regardless of why Shelley hid his name and later attributed to Mary the authorship of Frankenstein, the reality is obvious: Shelley, writing a novel with so many autobiographical details, would attract more attention to himself than to his work, and this was simply not in keeping with Shelley as a writer. It was keeping with Shelley's pattern to write controversial literature and veil himself as its author.[29] The implications of Shelley's views on alchemy, revolution by a new science, human relationships, human nature, and gender, within *Frankenstein* required a diversion, even if it meant he would never be credited for the novel.

The creature Shelley had in mind was within his nature already, as it was (from his point of view) the very nature of all mankind, though such a view would have been an indiscreet and religiously damning position to profess (as he found out by his expulsion from the University of Oxford for his tract on the *Necessity of Atheism*). In sympathy with Plato, Shelley's perfect man would appear physically and morally beautiful, however within 19th century England such a creature would be nothing less than *monstrous*. Shelley would have easily sympathized with a solitary, intelligent, compassionate, sexually androgynous creature; it was within Shelley already. However, to suggest that this perfect creature ought to be attempted, even realized as the uninhibited proto-type human, this unsettles the sensitivities of even the most open-minded person. Society's perspective of such a being already had been defined by its morals and religion: demon; deviant; monster; fiend!

In Plato's *Symposium* Shelley read of love, its origin among the first humans, and its destruction by division. He saw through the writing of Plato an image of the perfect person/man and also realized that this perfect original unity would be considered scandalous, monstrous, even demonic in his (and our) modern world. A careful reading of *Frankenstein* might well suggest that the monster with its gentle heart, its beneficent nature, its desire for love, is a recasting of what Plato describes in *The Symposium,* in a discourse from Asistophanes on Love. Although the passage is lengthy and much of the same was quoted in Chapter 1, the focus of the first chapter was upon *division* while the focus in the present chapter is upon *unity and love*. It is necessary that it is here again included in the main text; its importance is essential for approaching Shelley:

> You ought first to know the nature of man, and the adventures he has gone through; for his nature was anciently far different from that which it is at present. First, then, human beings were formerly not divided into two sexes, male and female; there was also a third, common to both the others, the name of which remains, though the

sex itself has disappeared. The androgynous sex, both in appearance and in name, was common both to male and female; its name alone remains, which labours under a reproach.

At the period to which I refer, the form of every human being was round, the back and the sides being circularly joined and each had four arms and as many legs; two faces fixed upon a round neck, exactly like each other; one head between the two faces; four ears, and two organs of generation; and everything else as from such proportions it is easy to conjecture. Man walked upright as now, in whatever direction he pleased; and when he wished to go fast he made use of all his eight limbs, and proceeded in a rapid motion by rolling circularly round, - like tumblers, who, with their legs in the air, tumble round and round. We account for the production of three sexes by supposing that, at the beginning, the male was produced from the Sun, the female from the Earth and that sex which participated in both sexes, from the Moon, by reason of the androgynous nature of the moon. They were round, and their mode of proceeding was round, from the similarity which must needs subsist between them and their parent.

They were strong also, and had aspiring thoughts. They it was who levied war against the Gods; and what Homer writes concerning Ephialtus and Otus, that they sought to ascend heaven and dethrone the Gods, in reality relates to this primitive people ... Jupiter, with some difficulty having devised a scheme, at length spoke, 'I think,' said he, 'I have contrived a method by which we may, by rendering the human race more feeble, quell the insolence which they exercise, without proceeding to their utter destruction. I will cut each of them in half; and so they will at once be weaker and more useful on account of their numbers. They shall walk upright on two legs. If they show any more insolence, and will not keep quiet, I will cut them up in half again, so they shall go about hopping on one leg.'

... Every one of us is thus half of what may be properly termed a man, and like a *psetta* cut in two, is the imperfect portion of an entire whole, perpetually necessitated to seek the half belonging to him. Those who are a section of what was formerly one man and one woman, are lovers of the female sex ... those women who are a section of what in its unity contained two women, are not much attracted by the male sex, but have their inclinations principally engaged by their own. And the *Hetairistriae* (Lesbians) belong to this division. Those who are a section of what in the beginning was entirely male seek the society of males; ...

Such as I have described is ever an affectionate lover and a faithful friend, delighting in that which is in conformity with his own nature. ...The cause of this desire is, that according to our original nature, we were once entire. The desire and the pursuit of integrity and union is that which we all love.[30]

In this context, we must ask, What was it that Shelley had in mind with the monster, the fiend, the demon in *Frankenstein?*

It is now plausible, even sustainable as an argument, that Shelley's *new* man of science was be the *ancient* man of an original and forgotten creation, or the original Metaphor prior to the Tower of Babel Moment. It is the goal of the alchemist to accomplish such a creature. It is clear that Shelley was fascinated, as personified and projected into Victor Frankenstein, with such a being whose unity and oneness also meant a higher love, a truer archetypal love, a love that might indeed be possible to recover if the principles of alchemy and science were to join force. Would the benefit to humanity be the (re)-creation of a perfect undivided being, a highly intelligent, gentle, beautiful, and loving being, *or* would the world see a monster, a demon, a fiend, an unnatural form whose potential to destroy might usher in the end of the world?

Let us first ask whether there is evidence that Shelley took his ideas, his fantasies, his well-documented research, and attempted a self reflection; was it possible Shelley desired to merge his own androgynous self into a place of unity and its themes occur in his writing?

D. Shelley's Epipsychidion, Or, The Soul within the Soul.

*We shall become the same, we shall be one
Spirit within two frames, oh! Wherefore two?
One passion in twin-hearts, which grows and grew*
—The Epipsychidion; Lines 573-575

"On the subject of the soul," or *The Epipsychidion*, was one more of Shelley's poems or works so deeply personal that it was initially his request to not be identified as the author. On February 16, 1821 - approximately 1 year and 5 months before his untimely death at age 29 - Shelley sent the first, and now lost copy of *The Epipsychidion* to his publisher Charles Ollier. Approximately 200 copies of the poem were printed however upon Shelley's death Mr. Ollier informed Mary Wollstonecraft Shelley that the final wishes of her husband were that the poem should be suppressed. Consider the progression that takes place, for in Shelley's works it is critical that the reader identify why the poet or novelist veils himself. With regard to *The Epipsychidion* Shelley moves from desiring anonymity to wishing the poem to never be seen, a step beyond the *Frankenstein* hoax which moved from anonymity to false authorship. The personal revelations in his poem on the subject of the soul would surpass even what Shelley felt was possible to veil. Why? His note to the editor might explain his own concern that very few readers were ready to cross the same threshold that he had, thus leading to scandal, attacks, and misinterpretations.

> My song, I fear that thou wilt find but few who fitly shalt conceive
> thy reasoning, of such hard matter does thou entertain.[31]

Notice that Shelley, even before writing the first words, indicates his uncertainty that the poem will be understood or find "fit" readers. He goes a step further and comforts himself in the knowledge that those who not fit are simply "dull;" that is to say, unenlightened. This in itself clues the careful reader to see Shelley's commitment to Illuminati and Rosicrucian methods, even if to expound his personal philosophy in terms more like Plato.

> Whence, if by misadventure, chance should bring Thee [the poem] to
> base company (as chance may do), quite unaware of what thou dost
> contain, I prithee, comfort thy sweet self again, my last delight! Tell
> them that they are dull, and bid them own that thou art beautiful.[32]

Keeping in mind the origin of love as described by Plato in *The Banquet (Symposium)* and the great likelihood that Shelley's (and Victor Frankenstein's) own alchemical agenda was the possibility that mankind might learn a means to rise and return to a higher place, a place of perfect unity and love without division, it is not surprising to discover that Shelley would wax poetically about his own love in these fragments to a longer poem – generally attached to the *Epipsychidion*. Whether it be the love of another or a reflection of himself which Shelley sees as the image of love, the alchemical creation is described as if taken from the writings of Plato and given life through the powers of an Agrippa, Paracelsus, or Albertus Magnus!

First, in the *Epipsychidion,* Shelley lays a foundation for his philosophy, Nature, that which yearns for unity; next, Shelley describes mankind which was undivided in the beginning and only by division (evil or Satan) does man fall from his undivided place where love is found:

Why there is first the God in heaven above, -
Who wrote a book called Nature, 'tis to be
Reviewed, I hear, in the next Quarterly;
And Socrates, the Jesus Christ of Greece,
And Jesus Christ Himself, did never cease
To urge all living things to love each other,
And to forgive their mutual faults, and smother
The Devil of disunion in their souls.[33]

Having set the groundwork for his world view, Shelley moves the reader into the uncomfortable position of being a spectator or eavesdropping disciple of the author's own experience:

I love you! – Listen, O embodied Ray
Of the great Brightness; I must pass away
While you remain, and these light words must be
Tokens by which you may remember me.
Start not – the ***thing* you are** is **unbetrayed,**
If you are human, and if but the shade
Of some sublimer spirit

And as to friend or mistress, 'tis a form;
Perhaps I wish you were one. Some declare
You are a familiar spirit, as you are;
Others with a ... more inhuman
Hint that, though not my wife, you are a woman;
What is the colour of your eyes and hair?
Why, *if you were a lady, it were fair*
The world should know – but, as I am afraid,
The Quarterly would bait you if betrayed;
And if, as it will be sport to see them stumble
Over all sorts of scandals. Hear them mumble
Their litany of **curses** – some guess right,
And **others swear you're a Hermaphrodite;**
Like that **sweet** marble **monster of both sexes**,
Which looks so sweet and gentle that **it vexes**
The very soul that the soul is gone
Which lifted from her limbs the veil of stone.[34]

Is it any wonder that Shelley first sought to publish the poem anonymously and later thought better of even the remote possibility that it could be traced to him and therefore chose to suppress it altogether?

The implications are either that Shelley has in mind an actual lover who is *so* perfect and *so unlike* anything human, that this *being* can only be described as a perfect unity, a return of Plato's original humanity; *or* Shelley is reflecting *within himself* and sees the unity of his own person in the image of the eternal One, neither male nor female, though existing as the undivided unity of masculine and feminine principles - in other words, he is the archetypal human, an androgynous and fully enlightened human. And what is the eschatological moment of such an elevated *love* or elevated *being*? According to Shelley (indebted to Greek ideals in Plato):

> And we will move possessing and possessed
> Wherever beauty on the earth's bare breast
> Lies like the shadow of thy soul – till we
> Become one being with the world we see[35]

Shelly is unmistakably declaring his fraternal link to the philosophy declared by the Greeks, the attempted creations of the alchemists (or actually realized creations, if one can believe some accounts), and to the ancient lore told and re-told before philosophers and alchemists in the sacred stories that became the foundation for all non-monotheistic religions. Shelley's personal and literary monster is only to be a reality when science mends the unnatural schism of the human, restores humanity to its undivided wholeness, and in an alchemical miracle, the androgynous *man* climbs back up the fragmented tower of his primordial past.

The question that lies before us now is the same which Shelley's monster felt within itself, *is there a place in the present world for such a being?* Is it to be a demon unleashed by our modern genetic alchemists or is it a return to the mythical garden? Either way, the serpent winds its tail around the Tree of Life and the path back to paradise is carpeted with the skeletons of would-be Creators chasing after their next androgynous Adam.

E. The Picture of Oscar Wilde

After a few minutes he became absorbed. It was the strangest book that he had ever read ...there were in it metaphors as monstrous as orchids, and as subtle in color. The life of the senses was described in the terms of mystical philosophy. One hardly knew at times whether one was reading the spiritual ecstasies of some Medieval

saint or the morbid confessions of a modern sinner. It was a poison-
ous book.

—Oscar Wilde, *The Picture of Dorian Gray*[36]

On Wednesday, 3rd April, 1895, Oscar Wilde stood on trial, or it might be more accurate to say that his *ideas* rather than his person were put on trial. During the afternoon session, Wilde was cross examined as to what readers might have understood concerning the morality behind *The Picture of Dorian Gray*. When asked by Edward Carson, the prosecutor, whether his novel might be interpreted in a manner that would lead to the corruption of Victorian morals, Wilde responds, "… you cannot ask me what misinterpretation of my work the ignorant, the illiterate, the foolish may put on it. It doesn't concern me. What concerns me in my art is my view and my feeling and why I made it; I don't care twopence what other people think about it."[37]

Carson's effort to trap Wilde into admitting that his art imitated his life would prove as difficult as herding cats, challenging but not impossible. Wilde was elusive, quick witted, and understood the implications quite well. If Carson could prove an association between Wilde's writing and his presumed immoral lifestyle, Wilde would be locked away for "acts of gross indecency." The evidence against Wilde was insurmountable and on the 25th May, 1895 Wilde was sentenced for two years hard labour.

If, as Carson argued, the life of the artist is reflected in the art, what was it that was staring back from *the picture of Oscar Wilde*? Was it the horrific decaying face of a man weathered by the indiscretions of his past, such as Dorian Gray's picture revealed, or was it something far deeper in Wilde than acts of gross indecency? We will answer and attempt to make a stronger case than Carson that it was something far deeper than moral indiscretions in Wilde that were carefully veiled, but evident nonetheless, in *The Picture of Dorian Gray*.

The preface of Wilde's only novel, *The Picture of Dorian Gray*, argued, "To reveal art and conceal the artist is art's aim."[38] Had the artist, or author in this case, violated his own principle and revealed himself in such a way that imprisonment was the unavoidable conclusion to the book? Are we to assume that the author was insightful enough to have established a principle concerning the morality or immorality of his own art and yet not see his own reflection staring back at him from the words on the page? In this case, the answer is given by Wilde in the preface, "It is the spectator, and not life, that art really mirrors."[39] In other words, Carson, *the spectator* of Wilde's art, saw a degenerate sodomite in Wilde which was more a reflection of Carson's inner demons than Wilde! Another reader, i.e., *spectator*, might see in the same novel a warning of the consequences of sin, the necessity of repentance, and find a deeply inspiring sermon, with Wilde as the Good Shepherd giving a metaphorical gothic homily for wandering narcissistic sheep. Which spectator is viewing the work accurately? Surely the maxim concerning the spectator of art bears consideration; a visitor to the Louvre might find a room filled with the statues of naked men, women, or a sleeping hermaphrodite as evidence of perversion in early Greek and Roman cultures, while another visitor admires the beauty of the human body and the skill of the sculptor. Holy or profane? According to Wilde, "those who find ugly meaning in beautiful things are corrupt without being charming. This is a fault. Those who find beautiful meanings in beautiful things are the cultivated. For these there is hope."[40]

But a higher question concerns us, what of those who find *meaning* beneath symbols? Clearly Wilde was writing in code, using symbols to reveal one thing to the enlightened while at the same time hiding his meaning to those who were not initiated into the mysteries of which he wrote. This is not an uncommon tactic, every great teacher from Sri Krishna to the Buddha, and Jesus Christ spoke in parables to enlighten their disciples and to confound their critics. We do not have to wonder how Wilde would answer our question, as he anticipates the question in his preface: "those who go beneath the surface do so at their peril. Those who read the symbol do so at their own peril."[41]

Our purpose is not to judge the man Wilde by *The Picture of Dorian Gray* nor is it to make moral judgments upon Wilde, rather our purpose is to direct the reader to consider what is undoubtedly a gaping hole in Wilde interpretation, particularly as it serves as one more critical piece of evidence in our present work. Over a century of literary analysis has flowed from the pens of scholars reading Wilde's masterpiece, yet none thus far have approached Wilde and *The Picture of Dorian Gray* within the context of fiction being used by an author as a means to advance an ancient alchemical agenda. We accept Wilde's "perilous" warning and recognize that while it is meant to keep the curious away, it is crucial to read Wilde beneath the symbol. We also recognize that the warning would be meaningless except for the fact that there is more beneath the surface and his alchemistic symbols are filled with meaning.

F. AUTOBIOGRAPHY OF AN ALCHEMIST

The Picture of Dorian Gray is a tale of a young man, unacquainted with the ways of the world, who is thrown into an alchemical cauldron by the hedonistic philosophy of Lord Henry Wotton. The naïve Dorian stands as a blank canvas, a *tabula rasa,* an unlearned young man, the perfect specimen for a sinister philosopher such as Lord Henry whose goal is to overturn the entire moral, philosophical, and theological structure of Victorian society by the creation of a perfect man. Like Victor Frankenstein's vision for his own new man, a body must be provided, a new mind must be developed, and a means must be found to effect the change, all of which is an alchemical process. Lord Henry has no need of electricity, his science is an ancient one, *words*, in particular the words associated with Greek philosophy:

> The aim of life is self-development. To realize one's nature perfectly – that is what each of us is here for. People are afraid of themselves nowadays. They have forgotten the highest of all duties, the duty that one owes to one's self... Courage has gone out of our race. Perhaps we never really had it. The terror of society, which is the basis of morals; the terror of God, which is the secret of religion – these are the two things that govern us ... and yet, I believe that if one man were to live out his life fully and completely, were to give form to every feeling, expression to every thought, reality to every dream – I believe that the world would gain such a fresh impulse of joy that we would forget all the maladies of medievalism, and return to the Hellenic ideal – to something finer, richer, than the Hellenic ideal, it may be. But the bravest man among us is afraid of himself. The mutilation of the savage has its tragic survival in the self-denial that mars our lives. We are punished for our refusals. Every impulse we strive to strangle broods in the mind and poisons us. The body sins once, and has done with its sin, for action is a mode of purification.[42]

Note, purification comes in the act of personal experience and not in the form of self sacrifice. The removal of the Church with its terrorizing God, the throwing off of traditional social mores with their binding laws, was to be replaced by the ideals of the Greeks, only taken one step further; this was Lord's Henry's laboratory and the specimen to be created was the young Dorian Gray.

> "Stop!" Faltered Dorian Gray. "Stop! You bewilder me" ... music had stirred him like that. Music had troubled him many times. But music was not articulate. It was not a new world, but rather another chaos, that it created in us. Words! Mere words! How terrible they were!... They seemed to be able to give a plastic form to formless things ...[43]

Dorian is entranced, hypnotically brought to a new stage of understanding by the philosophical prelude of Lord Henry's symphony to the self. What brought Dorian and Lord Henry together was an artist, Basil Hallward; another creator, or artistic alchemist, who would take a blank canvass – quite literally - and paint, or symbolize, the young Dorian in an ideal form; a portrait which would take a clueless formless human and unleash all of the potential, be it a god or demon. Basil put artistic form to a formless young man while Lord Henry took the same formless young man and put a new soul within him by preaching his philosophy of a new man. In the case of Lord Henry, the arrow did not miss the mark,

The *words* of Lord Henry were creative, giving form and matter to his formless intentions. Creation with a fire all its own, ideas in words! Lord Henry had only the need of sound, a frequency in harmony with the mind of Dorian to create his own new man. Dorian was in the process of dying to a previously empty life and at the same time being recreated and filled to a higher life. The sound of Lord Henry's words led Dorian to experience the ideal Greek life. The formless thing was breathing the ethereal higher life.

The same transformation, in form, was taking place on canvass as the artist infused his own passion, his soul, essentially his unspeakable *love* for the beautiful Dorian into the colored paints with a sort of spiritual intention. Basil explains the strange transformation that took place while painting Dorian as a transference of soul or life to an inanimate object,

> Every portrait that is painted with feeling is a portrait of the artist, not of the sitter. The sitter is merely the accident, the occasion. It is not he who is revealed by the painter; it is rather the painter who, on the colored canvas, reveals himself. The reason I will not exhibit the picture is that I am afraid that I have shown in it the secret of my own soul.[44]

And what exactly was the secret of Basil's soul? It was a forbidden emotion, an unspeakable passion, a *creative principle* arising from idealistic love. Such a secret buried within Basil was not only *un*natural but if expressed in certain visible forms, it was also illegal in Victorian England! Oscar Wilde found that even fiction was not a safe medium for expressing such love as his character Basil describes to Dorian in chapter 7 of *The Picture of Dorian Gray* as first published in the Lippincott magazine:

> It is quite true that I have worshipped you with far more romance of feeling than a man usually gives to a friend. Somehow, I had never loved a woman…Well from the moment I met you, your personality had the most extraordinary influence over me. I quite admit that I adored you madly, extravagantly, absurdly. I was jealous of every one to whom you spoke. I wanted to have you all to myself. I was

only happy when I was with you… One day I determined to paint a wonderful portrait of you. It was to have been my masterpiece. It is my masterpiece. But, as I worked at it, every flake and film of color seemed to me to reveal my secret. I grew afraid that the world would know my idolatry… You must not be angry with me, Dorian, for what I have told you. As I said to Harry once, you are made to be worshipped.[45]

It is an alarming concept, that the image of a perfect, albeit lifeless representation, on canvass, should become - through the intention of love - the creative principle uniting the soul of its creator with the true image itself, Dorian. The ideal, the perfect painted form or sacramental sign would *also* become the efficient instrument to give life to the soulless or formless Dorian in the very studio where the sacramental alchemy took place. Two creators – one contributing a hedonistic soul, another contributing a transforming image through love – worked their alchemy on a young formless subject. At the conclusion of the story, Dorian would reveal the new creature to his artistic creator Basil Hallward.

"You shall see it yourself tonight!" he cried, seizing a lamp from the table. "Come, it is your handiwork. Why shouldn't you look at it? You can tell the world all about it afterward if you choose. Nobody would believe you. If they did believe you, they would like me all the better for it. I know the age better than you do … "Yes," he continued, coming closer to him, and looking steadfastly into his stern eyes, "I shall show you my soul. You shall see the thing that you fancy only God can see."[46]

So it was that in words and by the instrument of a perfect image, a man with no previous life was born after the image of its creators. Dorian Gray became what Lord Henry and Basil Hallward infused into that formless thing that was once a young man. The scandal of the story is *not* the sins that are committed by the new creature who passes as Dorian Gray, but rather that such a being could be conceived, created, transposed as if from one substance or form to another! That the outward form or substance might appear the same while the internal essence or nature of that thing had been exchanged with another. With that, we are once again looking at the mediaeval doctrine of transubstantiation, fully realized as an alchemical doctrine for the transformation of man.

Indeed, the resemblance to the familiar formula for transforming or transubstantiating or effecting the alchemical effect, is all too palpable: An administrator or operator, i.e. priest is requisite; an appropriate instrument, form, or substance must be had; and proper words *and* intention must be applied. Unquestionably, Lord Henry stands in as the chief celebrant or creator or alchemist, with all the right words and intention to transform Dorian from one nature to another,

> ... the thought brought a gleam of pleasure into his [Lord Henry's] brown agate eyes – that it was through certain words of his, musical words said with musical utterance, that Dorian Gray's soul had turned ... to a large extent, the lad was his own creation. He had made him premature. That was something. Ordinary people waited till life disclosed to them its secrets, but to the few, to the elect, the mysteries of life were revealed before the veil was drawn away... Yes, the lad was premature. He was gathering his harvest while it was yet spring. The pulse and passion of youth were in him, but he was becoming self conscious. It was delightful to watch him. With his beautiful face and his beautiful soul, he was a thing to wonder at. It was no matter how it all ended, or was destined to end.[47]

The mysterious knowledge, the secrets, the creation of a new man? Victor Frankenstein would have been proud to stand with Lord Henry Wotton. The only missing piece to weave *The Picture of Dorian Gray* and *Frankenstein* into one alchemical tapestry of 19th century literature is a direct reference to alchemy as *intention*. As one might expect if Wilde was, like Shelley, framing his story behind an alchemical agenda, the evidence would be found in the characters words and actions.

> Soul and body, body and soul – how mysterious they were! There was animalism in the soul, and the body had its moments of spirituality. The senses could refine, and the intellect could degrade. Who could say where the fleshly impulse ceased, or the psychical impulse began? How shallow were the arbitrary definitions of ordinary psychologists! And yet how difficult to decide between the claims of the various schools! Was the shadow seated in the house of sin? Or was the body really in the soul, as Giordano Bruno thought? The separation of spirit from matter was a mystery, and the union of spirit with matter was a mystery also.[48]

Wilde's passing reference to Giordano Bruno, through the reflections of Lord Henry while conducting his transformation upon Dorian Gray, are telling.

Bruno (1548-1600) was a condemned heretic whose writings and theories included alchemical transformations of consciousness in the form of his art of memory.[49] As with any alchemist, the ultimate goal is the elevation of not simply one man or woman, but civilization itself, transformed by fire and ascending back to the place of unity, oneness, perfection. As the monster, the new man, the alchemical Dorian rises to his new place of infamy – for what else would the ordinary person in society see except a monster – he sees the alchemical vision for all things in him and through him!

> Indeed, there were many, especially among the very young men, who saw or fancied that they saw, in Dorian Gray the true realization of a type of which they had often dreamed in Eton or Oxford days – a type that was to combine something of the real culture of the scholar with all the grace and distinction and perfect manner of a citizen of the world. To them he seemed to be of the company of whom Dante describes as having sought to "make themselves perfect by the worship of beauty." Like Gautier, he was one for whom "the visible world existed."[50]

Note the description of Dorian as *type* or *perfect* or *one for whom the visible world existed; in short, he is a microcosm.* The elevated position, the recognized perfected form, the magnetic attraction that both men and women felt towards Dorian was scandalous preciously because it was in the form of one such as Dorian; a being unlike any others, an ageless and unnatural type of beauty that made men and women felt incapacitated in his presence. Dorian was to be desired or imitated in all ways, and he knew it and abused the power of such a gift from the creator,

> For while he was but too ready to accept the position that was almost immediately offered to him on his coming of age, and found, indeed, a subtle pleasure in the thought that he might really become to the London of his own day what to imperial Neronian Rome the author of the "Satyricon" once had been, yet in his inmost heart he desired to be something more … he sought to elaborate some new scheme of life that would have its reasoned philosophy and its ordered principles and find in the spiritualizing of the senses its highest realization… Yes there was to be, as Lord Henry had prophesied, a new Hedonism that was to recreate life, and to save it from the harsh, uncomely Puritanism that is having, in our own day, its curious revival…It may be, that our eyelids might open some morning upon a world that had

been refashioned anew in the darkness for our pleasure, a world in which things would have fresh shapes and colours, and be changed, or have other secrets, a world in which the past would have little or no place, or survive … *it was the creation of such worlds as these that seemed to Dorian Gray to be the true object, or among the true objects of life.*[51]

Wilde is not veiling a sodomitical agenda between the lines of *Dorian Gray*, rather he is expounding a vision, an alchemistic Utopia, and lest it be missed as something less, Wilde wrote his own commentary to the novel, again, in his own subtle way. In Wilde's essay, *The Critic as Artist*, published in July 1890, the same month and year that *The Picture of Dorian Gray* was published in Lippincott's Monthly Magazine, a rare glimpse into the alchemistic philosophy behind *Dorian Gray* is offered. Although Wilde writes his essay in the form of a dialogue, it reveals his agenda for a new society, one designed around the enlightened individual, an agenda that he cleverly hid in plain sight in *The Picture of Dorian Gray*. The creation of a new world, a new man (Dorian), was no mere piece of fiction for Wilde, and what worried him most was plainly stated, "I am but too conscious of the fact that we are born in an age when only the dull are treated seriously, and I live in terror of ***not being misunderstood***."[52] Indeed, Wilde ought to have lived in such terror for his revelations of such a scheme, as he acknowledged,

If we lived long enough to see the results of our actions it may be that …those whom the world calls evil [were] stirred by a noble joy. Each little thing that we do passes into the great machine of life which may **grind our virtues to powder** and make them worthless, *or **transform our sins into elements of a new civilization**, more marvelous and more splendid than any that has gone before.* [53]

The grinding of virtue to powder, transforming sins into elements of a new civilization, is unmistakable alchemistic language, for in some versions of the Philosophers' Stone, the Stone is referred to as a powder. It was bold and certainly unwise for Wilde to write an essay on purging the world of virtue, transforming society to a higher more splendid one through sin, and then publish such ideas in a serious essay during the same year that *The Picture of Dorian Gray* was published, analyzed, criticized, and condemned as obscene. And yet, he goes further than allusions to alchemy and a more splendid society, Wilde details how this great reversal or advancement should take place in a manner similar to a rite of initiation. As if to make sure that certain readers will understand his intention in writing Dorian Gray, he gives a commentary which easily explains what was happening between Dorian and Basil:

> By presenting high and worthy objects for the exercise of the emotions purifies and spiritualizes the man; nay, not merely does it spiritualize him, but it initiates him also into noble feelings of which he might else have known nothing, the word καθαρσις(katharsis) having, it has sometimes seemed to me, a definite allusion to the rite of initiation, if indeed that be not, as I am occasionally tempted to fancy, its true and only meaning here.[54]

Dorian is spiritualized by Lord Henry and by Basil, initiated through the emotions and intentions of both men. A change, a transformation, *the* catharsis(καθαρσις), is part and parcel of what takes place within a man during such heightened emotional and ritualistic moments, it is inseparable from the initiation which leads to change. It is the secret to religion, it is also the secret to the brotherhoods which have initiated men into their fold throughout the centuries. Symbols, words, actions, elevated emotions in re-enactments, death and rebirth – the spiritualized man, transformed, and the instrument of a higher more splendid society.

This was no fairy tale or science fiction piece of literature for Oscar Wilde; aside from his Greek idealism, he was a well versed, initiated, dandified Free Mason and understood better than most what was meant by *initiation*, catharsis, ritual, and rebirth. It is telling that Wilde's commitment to Masonic principals was thorough, right down to the costume and accoutrements. The author of *Dorian Gray* knew every ritualistic initiation step, from the proper clothing to the hidden meaning behind its ceremony. Wilde's enthusiasm to fully immerse himself in the externals of Free Masonry is evidenced by two significant events: (1) his being disciplined by the University of Oxford while there as a student for extravagant spending associated with Free Masonry;[55] and (2) when Wilde toured America in 1882 delivering lectures on aestheticism, he dressed in the very costume of the Apollo Lodge (Free Mason), where he held his membership in Oxford.

The externals of Masonry were but the outward form of what he held deep within as a philosophy of life. Wilde's aestheticism was no bare empty symbol, rather it was a firm belief in a higher order where beauty and life converge. A sacrifice must be paid -whether it be a Hiram Abif or a Jesus Christ - and illumination must be granted to enter into the mysteries of the ages. Resurrection for Wilde is the new man of the age, the illuminated man, the beautiful man, … it is *Dorian Gray*. As an 18th degree Mason, Wilde performed the role of Raphael in strikingly similar words and actions as one might expect from Lord Henry Wotton, when Wilde declared to the new initiates, *"I come to conduct you from the depths of darkness and the Valley of the Shadow of death to the Mansions of light."*[56] The mansions Wilde had in mind were not heavenly nor was the darkness associated with the danger of sin. To understand the alchemical mind of Wilde, one must climb upwards through the experience of what the world calls evil, and as the experience increases so too does the light. Freedom and beauty, if finally arrived at, comes through sacrifice as the former world of darkness (i.e. religion, society, morals, law) is rejected and the new man reaches full potential in the light of the One – the ideal, beauty, perfection. Communion is reached, unity is achieved.

Oscar Wilde with some Masonic Regalia

The high alchemical moment in Wildean fiction occurs in Chapter 11 of *The Picture of Dorian Gray*. With sufficient background, it may now be more fully appreciated, for it is nothing short of being a 19th century autobiography of an alchemist. A thorough examination of all of the alchemical elements in Chapter 11 of Wilde's classic novel, the same novel that was used as evidence against him, would require a book unto itself. For the sake of brevity and to avoid an overkill on this theme, Chapter 11 may be summarized as containing Dorian's alchemical pilgrimage step by step, including: Dorian's discovery of the use of gem stones;[57] Catholic altar furnishings including censors, tabernacles, and vestments to be worn by the priest during the act of transubstantiation;[58] mysticism; the study of astronomy; Darwinism;[59] the use of incense and psychedelic drugs that work on the brain's memories or trances;[60] music and its effects on human nature, including classical (Chopin and Schubert),[61] Indian pipes, South American Native instruments, Aztec bells,[62] and not surprisingly Wilde includes the writings of Alchemists (Pierre de Boniface) who describe how the nature and form of men may be changed.[63] The conclusion which is reached by Dorian after such a thorough study into hermeticism and the use of sacred objects, is strangely familiar to the doctrines found in ancient Greek writings, particularly Plato, and what is described in the Vedic scriptures.

> To him, man was a being with myriad lives and myriad sensations, a complex multiform creature that bore within itself strange legacies of thought and passion, and whose very flesh was tainted with the monstrous maladies of the dead... Had some strange poisonous germ crept from body to body till it had reached his own? ... there were times when it appeared to Dorian Gray that the whole history was merely the record of his own life, not as he has lived it in act and circumstance, but as his imagination had created it for him, as it had been in his brain and in his passions.[64]

If *The Picture of Dorian Gray* was Wilde's expose of the alchemical meme of the transformation of man, then it is significant that one of the people mentioned in that crucial and highly alchemical eleventh chapter is Dante Alighieri, and a brief view of Dante's climb out of the pit is in order.

G. Dante's Pit and Climb Out of Hell

In what must surely be the world's most famous mid-life crisis-in-poetry, Dante Alighieri begins his *Divine Comedy* with some sobering words concerning Christianity's doctrine of the inescapable hell:

I AM THE WAY INTO THE CITY OF WOE,
I AM THE WAY TO A FORSAKEN PEOPLE.
I AM THE WAY INTO ETERNAL SORROW.

SACRED JUSTICE MOVED MY ARCHITECT.
I WAS RAISED HERE BY DIVINE OMNIPOTENCE,
PRIMORDIAL LOVE AND ULTIMATE INTELLECT.

ONLY THOSE ELEMENTS TIME CANNOT WEAK
WERE MADE BEFORE ME, AND BEYOND TIME I STAND.
ABANDON HOPE ALL YE WHO ENTER HERE.[65]

There has been, in the course of Dante criticism, no end of commentary on the belief of critics that Dante's poem is a Christian work, deeply suffused with the doctrines of scholastic theology, and in particular, of Thomas Aquinas.

Yet, as we saw, there may have been something else at work in Aquinas, and that means there may have been something else at work in Dante, and once it again, it was Oscar Wilde who first strongly suggested - in that magical and beautifully Baroque and highly alchemical eleventh chapter of *The Picture of Dorian Gray* - that something else entirely was at work in Dante's "worship of beauty," something suggested by the heavily esoteric chapter in which Wilde's remarks occur.[66]

This perspective on Dante should not, however, be too surprising, for when viewed objectively, nothing in Dante's *Divine Comedy* can be viewed as remotely close to orthodox Christianity, for Dante's guide through Hell - Virgil - is a pagan, and Dante manages to escape that Hell, itself an impossibility in orthodox Catholic dogma, by climbing up the back of Satan himself, i.e., *by using the devil to escape evil and to lead him to the ultimate transformation of the beatific vision*, a vision in which all distinctions are again united in a transcendent expression and union of love.

This fantastic alchemical and Gnostic inversion was not lost upon occultists, for the famous Eliphas Levi, in his *The History of Magic*, commented at length on Dante's "Christian" poem:

Amidst a great multiplicity of commentaries and studies on the work of Dante, no one, that we are aware, has signalised its characteristic-in-chief. The masterpiece of the glorious Ghibelline is a declaration of war against the papacy by a daring revelation of mysteries. The epic of Dante is Johannite and Gnostic; it is a bold application of Kabalistic figures and numbers to Christian dogmas, and is further a secret negation of the absolute element there; his visit to the supernatural worlds takes places like an initiation into the Mysteries of Eleusis and Thebes. He is guided and protected by Virgil amidst the circles of the new Tartarus, as if the tender and melancholy prophet of the destinies of the son of Pollio were, in the eyes of the Florentine poet, the illegitimate yet true father of the Christian epic. Thanks to the pagan genius of Virgil, Dante emerges from that gulf above the door of which he had read the sentence of despair; he escapes by standing on his head, which means by reversing dogma. So does he ascend to the light, using the demon himself, like a monstrous ladder; by the force of terror he emerges from terror, from the horrible by the power of horror. He seems to testify that hell is without egress for those only who cannot go back on themselves; he takes the devil against the grain, if I may use so familiar an expression, and attains emancipation by audacity. This is truly protestantism(sic) surpassed, and the poet of Rome's enemies has already divined Faust ascending to heaven on the head of de-feated Mephistopheles. Observe also that the hell of Dante is but a negative purgatory, by which is meant that his purgatory seems to take form in his hell, as if in a mould; it is like the lid or stopper of the gulf, and it will be understood that the Florentine titan in scal-ing Paradise meant to kick purgatory into hell.

His heaven is composed of a series of Kabalistic circles divided by a cross, like the pentacle of Ezekiel; in the centre of this cross a rose blossoms, thus for the first time manifesting publicly and almost ex-plaining categorically the symbol of the Rosicrucians. We say for the first time because WIlliam of Lorris, who died in 1260, five years be-fore the birth of Dante, did not complete the *Romance of the Rose*, his mantle falling upon Clopinel some fifty years later. It will be discov-ered with a certain astonishment that the *Romance of the Rose* and the *Divine Comedy* are two opposite forms of a single work - initiation by independence of spirit; satire on all contemporary institutions and an allegorical formula of the grand secrets of the Brotherhood of the Rosy Cross.[67]

Indeed, Levi does not stop there, but goes on to suggest that Dante's inver-sions were perhaps the agenda of a hidden network:

Whilst religious wars incarnardined the world, secret illuministic associations, which were nothing but theurgic and magical schools, were incorporated in Germany. The most ancient of these seems to have been that of the Rosicrucians, whose symbols go back to the times of the Geuplhs and Ghibellines, as we see by the allegories in the poem of Dante and by the emblems in the *Romance of the Rose.*"[68]

But why, or how, would Wilde have picked up on such notions of Dante, and hint at them in his remarkable novel?

The answer is very simple: Wilde, as we saw, was a Freemason, and it is therefore to Masonry that we must repair to understand the role that the symbols of androgyny has in the initiation rites of the world's most famous secret society. For there we will find even further literary reversals of the Tower of Babel moment of History in an initiatory symbolization of androgynous "alchemosexuality."

ENDNOTES

1   The authors are aware that Mary Shelley has been credited with the authorship of Frankenstein although partial credit as co-author has more recently been given to Percy Shelley. It is the position of the authors, based primarily on textual evidence with a careful reading of extra textual evidence, that Percy Shelley ought to be viewed the *sole* author of the 1818 first edition of Frankenstein. The authors also hold to the position that Percy Bysshe Shelley intentionally perpetrated the Mary Shelley authorship hoax *himself*, with his wife's and friends' full knowledge. Throughout this chapter when the name Shelley is used with regard to authorship, the authors are indicating Percy Shelley. It is believed that the Shelley authorship of Frankenstein is key to interpreting underlying themes within the novel, however for the sake of brevity in this chapter and book, the argument over authorship will be minimal and left for the reader to independently research at greater length if interested. A recommended starting point for readers inclined towards researching textual and extra textual material related to authorship is a monumental work by Phyllis Zimmerman, *Shelley's Fiction*, Darami Press, 1998; and also John Lauritsen's *The Man Who Wrote Frankenstein*, Pagan Press, 2007. In terms of Shelley's authorship we are in full agreement with both works, however our analysis as to what conclusions might be drawn from Shelley's authorship are original and should neither be considered dependent upon nor a reflection of Zimmerman's or Lauritsen's valuable contributions to this neglected issue.

2   N.B., In Edward Ellerker William's journal written during his last days with Shelley, William's gives very clear evidence that Shelley's method for *writing* was to regularly use an amanuenses. Shelley would dictate while another would "write." It is not difficult to see how an obvious confusion would arise if Mary Wollstonecraft Shelley's journals or letters mention that she "wrote" Frankenstein, particularly when she *did write* as Shelley dictated and later edited, corrected, revised what had been *written* by Mary. Granted it becomes more confusing to untangle the authorship conspiracy considering that Shelley intentionally wanted his name unknown while later promoting

Mary as the author. It is also evident in reading Williams' journal that if Shelley wanted to gain a wider audience on a work and was well aware from his earliest writings that his name would detract from the work or call attention to himself where it was unwanted, he thought nothing of using a nom de plume or allowing another name to receive the attention for a work of his own, cf. Williams journal entry:

Sunday, November 11th 1821. "In the evening Shelley proposes to me to assist him in a continuation of the translation of Spinoza's Theologico-political tract, to which **Lord Byron has consented to put his name**, and to give it greater currency, will write the life of that celebrated Jew to preface the work."

Monday, November 12th. Shelley and I commenced Spinoza, that is to say, **I write** while *he dictates*. Write from page 178-188."

It is known that Shelley wrote under a pseudonym as well, ironically one such pseudonym was "Victor" as in his poems published with his sister Elizabeth (another key character in Frankenstein), *Poetry by Victor and Cazire*.

3   Mary Shelley (with Percy Shelley), *The Original Frankenstein*, Edited by Charles E. Robinson, Vintage Books:New York, 2008, p.189

4   Mary Shelley (with Percy Shelley), *The Original Frankenstein*, Edited by Charles E. Robinson, Vintage Books:New York, 2008, pp. 440-441

5   Mary Shelley's *Introduction* to the 1831 edition of *Frankenstein.*

6   *The Original Frankenstein*, Edited by Charles E. Robinson, p. 441

7   In the course of researching for the present book it became evident to the authors that such a book is a critical missing component to modern issues in science, religion, and alchemy. Subsequently, a book on the topic of Shelley's life as it relates to such issues is presently being undertaken by Scott de Hart.

8   One description of Shelley while at Eton notes, "[Shelley] the scorner of games and muscular amusements, could not hope to find much favour with such martinets of juvenile convention as a public school is wont to breed." Symonds supplies no shortage of evidence and first hand accounts which might be added from Shelley's childhood to further support the isolation Shelley felt in part due to his feminine features, his lack of interest in athletics, his gentleness, and advanced intelligence, hardly characteristics then (as well as today) which elevate a young man to popularity among his peers.
     John Addington Symonds, *Percy Bysshe Shelley*, Project Gutenberg, EBook posting date: August 14, 2009, #4555, p.41 (originally published 1878)

9   Ibid, pp.98-99.

10   Ibid p.26

11   Op. cit. James Bieri, *Percy Bysshe Shelley, A Biography*, John Hopkins University Press:Baltimore, 2008, pp.57-60

12   Mary Shelley (with Percy Shelley), *The Original Frankenstein*, Edited by Charles E. Robinson, Vintage Books:New York, 2008, pp.63-64

13   John Addington Symonds, *Percy Bysshe Shelley*, Project Gutenberg, EBook posting date: August 14, 2009, #4555, p.41 (originally published 1878), pp.55-56

14   Mary Shelley (with Percy Shelley), *The Original Frankenstein*, Edited by Charles E. Robinson, p. 176

15   Ibid, pp.158-159

16   Ibid, p.172

17   Aristophanes origin of love as described in Plato's Symposium.

18   Lord Byron, the same poet who was with Shelley during the Swiss summer in which Mary Shelley describes the wager on which Frankenstein was first thought of, was notorious for his relationships with both men and women. Another companion with Shelley and Byron at the same time was John Polidori, Byron's handsome young physician. Thomas Jefferson Hogg, perhaps Shelley's closest friend was known for his attempted (and encouraged) sexual relationships with Shelley's first wife Harriet and later with Mary Wollstonecraft Shelley, all three agreeing on the ideal of 'free love.' Shelley was also especially close to Edward John Trelawny, Thomas Medwin, and Edward Ellerker

Williams, with whom he died in the boating accident, and with whom it is even possible that his ashes were mingled after cremation. The nature of these relationships has not been overlooked by Shelley biographers, most agreeing that the friendships were intimate, though no *explicit* proof is given that such relationships were sexual.

19 Shelley, *Frankenstein*, p. 212

20 Quoted from John Lauritsen, *Hellenism and Homoeroticism in Shelley and his Circle*, 2008. www.paganpressbooks.com/jpl/HHREV3.htm

21 Shelley, *Frankenstein*, p.48

22 Ibid, p.55

23 Note, among Shelley's earliest writings was *St. Irvyne or the Rosicrucian*; a sympathy with Rosicrucian ideas and the founding of his own Illuminati styled group are underlying themes in his own desire to reform society through secret knowledge. A later work, begun near the same period as *Frankenstein*, was an unfinished story titled *The Assassins*, a Gnostic styled group with Illuminati characteristics. The fact that Shelley places *Frankenstein* in Ingolstadt, the home of the Illuminati, and dares not reveal more than the location as it associates to Victor's research, study, and creation, is a key point for consideration.

24 Shelley, *Frankenstein*, p. 168

25 Plato is described as "one of Shelley's favorite authors." "The mixture of the poet and the sage in Plato fascinated him. The doctrine of anamnesis, which offers so strange a vista to speculative reverie ... took a strong hold upon his imagination." Shelley's study of Plato's *Dialogues*, which he would later translate, "acted powerfully on the poet's sympathetic intellect" according to his friend Thomas Jefferson Hogg. (Symonds, *Percy Bysshe Shelley*, p.82)

26 Shelley, *Frankenstein*, p. 168

27 Shelley, *Discourse on the Manners of the Ancient Greeks*. Published as Plato: *The Banquet*, translated by Percy Bysshe Shelley, forward by John Lauritsen, Pagan Press, 2001. pp.16-17

28 Plato: *The Banquet*, translated by Percy Bysshe Shelley, forward by John Lauritsen, Pagan Press, 2001. p.7

29 A book examining Shelley's pattern to write controversial literature and avoid identification with it, particularly in relationship to how Frankenstein fits Shelley's agenda, is presently being written by the author of the present work.

30 Plato: *The Banquet*, translated by Percy Bysshe Shelley, forward by John Lauritsen, Pagan Press, 2001. pp.47-51

31 Percy Bysshe Shelley, *The Complete Poetical Works*, Oxford Edition, 1914, Edited by Thomas Hutchinson. (E-book) p. 2091

32 Ibid, p. 2092

33 Ibid, pp. 1918-1919

34 Ibid, pp. 1920-1921

35 Ibid, p. 1932

36 Oscar Wilde, *The Picture of Dorian Gray*, Barnes and Noble Classic:New York, 2003. p.129

37 Merlin Holland, *The Real Trial of Oscar Wilde*, Perennial: New York, 2004, p.81

38 Wilde, *The Picture of Dorian Gray*, p.1

39 Ibid, p.2

40 Ibid, p.1

41 Ibid, p.2

42 Ibid, pp. 20-21

43 Ibid, p.21

44 Ibid, p.7

45 Oscar Wilde, *The Picture of Dorian Gray*, Lippincott Edition. New York, 1988, p.250

46 Wilde, *The Picture of Dorian Gray*, Barnes and Noble Edition, 2003, pp. 156-157

47 Ibid, pp. 61-62

48 Ibid, p. 62

49 For the whole connection of Giordano Bruno to the Renaissance hermetic tradition see Frances A. Yates, *Giordano Bruno and the Hermetic Tradition*, Volume II, *Frances Yates: Selected Works* (Routledge, 2001) and Frances A. Yates, *The Art of Memory*, Volume III, *Frances Yates: Selected Works* (Routledge, 2001).

50 Ibid, p. 132

51 Ibid, pp. 132-135

52 Oscar Wilde, *The Complete Works of Oscar Wilde*, (London, 2003), p. 1114

53 Ibid, essay, *The Critic as Artist*, p.1121

54 Ibid, p.1117

55 On three separate occasions, on 8th, 22nd November 187 and on 22nd May   1878, Wilde was summoned before the University Chancellor's Court where action was brought against him for non-payment of outstanding debts. The second of these summonses is of direct relevance because it entailed the purchase of Masonic regalia. In November of 1876 he spent a total of £15.18.6, a vast amount at the time, equivalent to some £650.00 in today's terms, to purchase from George Henry Osmond Watch and Clock Makers of 118 St. Aldate Oxford, various items which included: 18 carat gold and ivory studs, a lamb skin Rose Croix apron & collar, a Rose Croix jewel, sword and belt as well as a Masonic leather jewel case, lettered with his initials. Cf. freemasonry.com/beresiner8.html and Richard Ellmann's, *Oscar Wilde* (Vintage:1988), pp. 40, 68, 164

56 Richard Ellmann, *Oscar Wilde* (Vintage:1988), p. 68

57 Wilde, *The Picture of Dorian Gray*, pp. 138-140.

58 Ibid., pp. 135-136.

59 Ibid., p. 136.

60 Ibid., pp. 136-137.

61 Ibid., pp. 137-138.

62 Ibid.

63 Ibid., p. 139.

64 Ibid., pp. 147-148.

65 Dante Alighieri, *The Divine Comedy: Inferno*, Canto III, trans. John Ciardi (New American Library, 2003), pp. 30-31, emphasis in the original.

66 Oscar Wilde, *The Picture of Dorian Gray* (Barnes and Noble), p. 132.

67 Eliphas Levi, *The History of Magic*, Trans. A.E. Waite (Rider Books, 1986), pp. 260-261.

68 Ibid., p. 263.

❧ Ten ❧

THE ESOTERIC ANDROGYNY:
SECRET SOCIETIES AND THE HIDDEN TRADITIONS OF
ALCHEMOSEXUAL MAN

∵

"It influences the thoughts of those obscure prophetical writers, like Joachim of Flora, strange dreamers in a world of flowery rhetoric of that third and final dispensation of a 'spirit of freedom,' in which law shall have passed away."
—Walter Pater[1]

"You are now in this Degree permitted to extend your researches into the more hidden paths of nature and science."
—From the Initiation of the Fellow Craft[2]

A. ALCHEMOSEXUALITY IN MASONIC INITIATION
1. The First Degree: Entered Apprentice

IMAGINE, FOR A MOMENT, that you are standing in a darkened room upon a floor whose black and white tiles are laid out like a checkerboard, except they are all diagonal to the room's walls, and imagine that the entire process of initiations through various degrees are comprehended by you in one moment. Around your neck, there is a noose, called in the parlance a "cable-tow." You have been made to strip off your shirt on your left side, exposing your left breast, to roll up your left pant leg, take your shoes off, put a slipper on your right foot, and wear a blindfold, called a "hoodwink."[3] Then, the needle of a geometer's compass is pressed to your left breast, and the following words are uttered:

Mr. (N.), on entering this Lodge for the first time, I receive you on the point of a sharp instrument pressing your naked left breast, which is to teach you, as it is a torture to your flesh, so should the recollection of it ever be to your mind and conscience, should you attempt to reveal the secrets of Masonry unlawfully.[4]

No females are present. After the compass needle has been taken away, you are eventually led to an altar, made to kneel on your left knee with your right leg bent at the knee at a right angle - still blindfolded - placing your left hand beneath a book, which, it turns out, will be a Bible, and your right hand on top of it, touching a compass and square. A prayer is then said, and then you are made to swear an oath, which includes this:

All this I most solemnly, sincerely promise and swear, with a firm and steadfast resolution to perform the same, without any mental reservation or secret evasion of mind whatever, binding myself under no less penalty than that of having my throat cut across, my tongue torn out by its roots, and *my body buried in the rough sands of the sea*, at low-water mark, where the tide ebbs and flows twice in twenty-four hours, should I eve knowingly violate my Entered Apprentice obligation. So help me God, and keep me steadfast in the due performance of the same.[5]

At this juncture, the words of Genesis 1:1-2, are recited - "In the beginning God created the heaveans and the earth. And the earth was without form, and void,and darkness was upon the face of the waters. And God said, Let there be light, and there was light." - and the blindfold is removed, and the room is fully illuminated.[6] You have just been illumined, and are now an Entered Apprentice, and all of it has taken place utterly devoid of any female presence, for this mildly-alchemosexual ritual has just admitted you to one of the world's oldest, best-known, and secret fraternities, the Freemasons. You are now one of many brothers, all of whom have undergone the same vague alchemosexual liturgy. And throughout your progress through each of Masonry's succeeding degrees, all of them, without exception, will be accomplished by, surrounded by, and done for, the brothers. No females allowed. The ritual of the first degree revolves around two inescapable focal points: men, and creation.

This vaguely alchemosexual context is more explicitly elaborated by the great Masonic expositor, Albert Pike in his Scottish Rite "bible," *Morals and Dogma:*

Remembering what we have already said elsewhere in regard to the old ideas concerning the Deity, and repeating it as little as possible, let us once more put ourselves in communion with the Ancient poetic and philosophic mind, and endeavor to learn of it what it thought, and how it solved the great problems that have ever tortured the human intellect.

The division of the First and Supreme Cause into two parts, one Active and the other Passive, the Universe Agent and Patient, or the hermaphroditic God-World, is one of the most ancient and widespread dogmas of philosophy or natural theology. Almost every ancient people gave it a place in their worship, their mysteries, and their ceremonies.[7]

Pike elsewhere notes that Active and Passive causes are Male and Female, respectively,[8] and thus, God or the physical medium, can be referred to as

the Grand Whole, or the single hermaphroditic Being that comprehends all existences..."[9]

But what has all this to do with the initiation of an Entered Apprentice, or the first degree of Blue Lodge Masonry?

Quite a bit, as it turns out, for it will be recalled that the prayer and oath occur at an altar, on which rests a Bible, *over* which is placed the Masonic implements of the Compass and Square. After the creation verses of Genesis 1:1-2 have been recited and the "hoodwink" removed," the initiate sees the display on the altar. Pike comments on the symbolism here as follows:

The Hermaphroditic figure is the Symbol of the double nature anciently assigned to the Deity, as Generator and Producer, as the BRAHM and MAYA among the Aryans, Osiris and Isis among the Egyptians. As the Sun was male, so the Moon was female; and Isis was both the sister and the wife of Osiris. *The Compass, therefore, is the Hermetic Symbol of the Creative Deity*, and the Square of the Productive Earth or Universe.... *The Compass, therefore, as the Symbol of the Heavens, represents the spiritual, intellectual, and moral portion of this double nature of Humanity...*[10]

With Pike's reference to the Compass representing "this double nature of Humanity" the mask is off, and the alchemosexual imagery of the Compass its androgynous symbolism - is laid bare.

However, this places the symbolism evident in the Entered Apprentice degree of Masonry - at least as far as Pike's Scottish Rite Masonry is concerned - into a peculiar light, for consider once again the context: (1) a recitation of a creation myth, (2) in the context of an initiation that (3) is being enacted and enabled entirely by men. In other words, the symbolism suggested here is that Masonry has constituted itself as a parallel magisterium - a kind of "unapostolic succession" to that of the Church - with its own unique tradition of *interpretation* of an authoritative text, an interpretation that is Hermetic and alchemosexual in its very foundations. These alchemosexual overtones are not without their own social, aesthetic, and ethical consequences, for it means that in the Masonic goal to form the "rough Ashlar" of imperfect human-ity into the "perfect Ashlar" of a tolerant and civil society, that an alchemo-sexual transformation of mankind's consciousness - both social, sexual, and individual - is an inevitable goal.[11] Just exactly how, we shall see in the next section on Oscar Wilde, but for the moment, it will be helpful to review the initiations of the second and third degrees of Freemasonry - the Fellow Craft and Master Mason - respectively.

2. The Second Degree: Fellow Craft and the Gradual Revelation of the Primordial Alchemosexuality

In the initiation to the Second Degree of Fellow Craft, the candidate is again prepared outside the lodge by rolling up his *right* pants leg, removing his shoes and socks, and stripping his shirt from his right arm and this time exposing the right breast. Again, the hoodwink or blindfold is placed over his eyes and the cable tow or noose is placed around his right arm where it meets the shoulder. He is given a slipper to wear on his left foot.[12] This time however, it is the Square which is removed from the altar, with the point of its right angle being pressed to the initiate's right breast.[13] Once inside the lodge, the candidate is again conducted to the altar, when he is made to kneel, this time upon his right exposed knee with his left leg bent at a right angle. His right hand rests upon the open Bible once again, upon which the Compass and Square are again placed, only this time, one point of the Compass is placed *above* the Square, over the left side of the Square as the candidate would see it (had he not been "hoodwinked"!).[14]

Again, an oath is sworn, prescribing horrible penalties that include tear-ing open the candidate's breast and having his heart plucked out - notice the sacrificial imagery of a ritual that would come to gruesome reality in the human sacrifices of Meso-America - and, in some cases, his vitals thrown over his left shoulder.[15] The twin pillars of the lodge, called Jachin and Boaz are then revealed,[16] and the candidate is then informed that by comtemplating them, "we are inspired with a due reverence for the Deity and his works and are induced to encourage the studies of astronomy, geography, navigation, and the arts dependent on them, by which society has been so much benefited."[17] The initiate is further informed that the original Joachin and Boaz pillars were constructed hollow and that they "contained the archives of Masonry."[18]

The initiate is then address by the "Worshipful Master" of the lodge, and here it is best to cite the ritual at length, for it contains numerous important clues and symbols for our purposes:

> I shall now direct your attention to the letter "G" (here the Master turns and points to a large gilded letter "G," which is generally placed on the wall back of the Master's seat and above his head; some Lodges suspend it in front of the Master, by a cord or wire), which is the initial of geometry, the fifth science, it being that on which the Degree was principally founded.
>
> *Geometry, the first and noblest of science, is the basis upon which the superstructure of Masonry is erected.* By geometry, we may curiously *trace nature through her various windings to her most concealed recesses.* By it we discover the power, the wisdom, and the goodness of the Grand Artificer of the Universe and view with delight *the proportions which connect this vast machine....*
>
> The lapse of time, the ruthless hand of ignorance, and the devastations of war have laid waste and destroyed many valuable monuments of antiquity on which the utmost exertions of human genius have been employed. Even the Temple of Solomon... escaped not the unsparing ravages of barbarous force. *Freemasonry, notwithstanding, has still survived....* Tools and implements of architecture are selected by the fraternity, to imprint on the memory wise and serious truths; and *thus, through a succession of ages, are transmitted unimpaired the excellent tenets of our institution.*[19]

Before we can comment on this lengthy passage, there are two more short admonitions that we must look at, given to the initiate in the "charge":

> Geometry, or Masonry, *originally synonymous terms, being of a divine and moral nature,* is enriched with the most useful knowledge; while it proves the wonderful properties of nature, *it demonstrates the more important truths of morality.*[20]

The second and final short passage with which we must be concerned is this. Pointing to the letter "G" again on the wall of the lodge, the initiated Fellow Craft is asked "To what does this allude?" The answer follows immediately:

Geometry, the fifth science; but more particularly to the sacred name of the Deity, to whom we should all...with reverence most devoutly and humbly bow.[21]

From the standpoint of a "topological" and "alchemosexual" metaphor preserved from High Antiquity, what does all this mean?

We may summarize the emphasized points of the above-quoted passages as follows:

1) Geometry is almost asserted to be synonymous with God, for God is viewed as the supreme Geometer, or, to put it in the parlance of the more-recently developed higher mathematical language of topology, the supreme Topologist;

2) Geometry - and by extension of the metaphor, topology - is explicitly stated to be synonymous with Masonry and the basis on which its whole superstructure is erected;

3) Geometry - and by extension of the metaphor, topology - is the basis on which to divine the secrets of nature *"through her various windings to her most concealed recesses;"* a curious image, for "windings" is of course a specifically topological term, and the imagery, we are bold to suggest, alludes to the memory of a profound and deep physics, namely, that of *rotation* or *torsion* as being a fundamental principle of differentiation in the physical medium, or "nature;"

4) Masonry, in this degree, explicitly asserts its belief that it is a continuous body of knowledge that has survived from antiquity. This fact, coupled with the idea of Geometry as not only the supreme "science" but even a metaphor for God, plus the fact of the placement of the Compass and Square *over* the Bible, suggests that Masonry views itself as a parallel magisterium to the Church, but with a very *different* interpretation of a "canonical text," an interpretation based upon its understanding of a geometric, that is to say, topological metaphor; and finally,

5) For our purposes, note that in this degree, one point of the Compass, which it will be recalled represents God the Geometer in His primordial alchemosexuality, is placed over the Square,

which symbolizes the purely feminine. In other words, symbolically, the process of initiation through the three degrees, interpreted from the standpoint of Pike's Scottish Rite *Morals and Dogma,* is a gradual initiation into a fully symbolized alchemosexuality, and only completely realized in the next degree, the degree of Master Mason, as we shall now see.

3. The Master Mason and The Full Revelation of Primordial Alchemosexuality in the Context of Geometry and the Topological Metaphor

We begin the explication of the Third Degree, that of Master Mason, with the symbol of the degree itself:

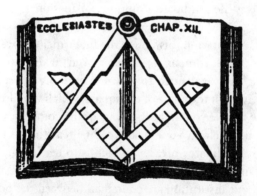

Third Degree Master Mason Symbolism[22]

Note in this symbolism, the Compass, symbol of the supreme Geometer or God, and also of His primordial alchemosexuality per Albert Pike, is now placed with *both* points over the Square, or, as Duncan suggestively states it, "both points of the compasses are *elevated above* the square."[23] Viewed in the context of our previous conclusions that Masonry views itself as a continuous, though hidden and parallel magisterium with the Church, with its own unique Masonic interpretation of a textual canon, this symbolism is highly significant, for it suggests that this geometric and alchemosexual metaphor will become *the* principal exegetical and interpretive principle, constituting an ideological and metaphysical first principle unique to "the brethren." It also suggests that Masonry knows of the direct connection between the topological metaphor, and alchemosexuality, but that, for whatever reason, it has obscured that deep relationship by a complex symbolism.

Such possibilities are even more strongly suggested by the ritual of initiating a Master Mason itself, for it is *this* ritual that is particularly associated, within Masonry, to resurrection and to illumination or enlightenment. Indeed, Duncan's *Ritual* suggests that all of these associations - resurrection, enlightenment, the topological metaphor, and the primordial alchemosexuality symbolized - make it "the height of Ancient Freemasonry, and the most sublime of all the degrees in Masonry (Royal Arch not even excepted)..."[24]

Again, the candidate for the degree is made to prepare outside the lodge, this time by rolling up *both* pants' legs, and removing his shirt entirely, exposing his naked chest. Again, he is hoodwinked, and the cable-tow or noose is now placed around his waste. Again, he removes his shoes but this time no slippers are to be worn on a single foot; both feet remain bare.[25]

Being met again at the door of the lodge, this time the candidate has *both* points of the Compass pressed to his left and right breasts. As this is done, the Senior Deacon gives the candidate a unique explanation:

Brother (N.), on entering this Lodge the first time, you were received on the point of the compasses, pressing your naked left breast, the moral of which was explained to you. On entering the second time, you were received on the angle of the square, which was also explained to you. I now receive you on both points of the compasses, extending from your naked left to your naked right breast(he he places both points against candidate's breasts), which is to teach you, that as the vital parts of man are contained within the breasts, so the most excellent tenets of our institution are contained between the points of the compasses - which are Friendship, Morality, and Brotherly Love.[26]

Again the candidate is led, still "hoodwinked," to the altar, this time kneeling on both knees, and placing both hands on the bible, square, and compass, which is laid upon the bible as depicted in the picture beginning this section.

After the hoodwink has been removed, the "Worshipful Master" again stresses the nature of this new illumination:

Brother (N.), on receiving further light, you perceive more than you have heretofore. Both points of the compasses are elevated above the square, which is to teach you never to lose sight of those truly Masonic virtues, which are friendship, morality, and brotherly love.[27]

Subsequently, as the ritual unfolds, the newly initiated Master Mason is shown, or given, a trowel - whose phallic symbolism should not be overlooked

- while the "Worshipful Master" informs him that "we, as Free and Accepted Masons are taught to *make use of it for the* more noble and *glorious purpose of spreading the cement* of brotherly love *and affection.*"[28] Here the wording explicitly suggests - almost too explicitly - an alchemosexual image, but now, in addition to that, *practice.* Certainly most Masons would balk and pale at such implications, but nonetheless, they are there, and follow by careful consideration of their own authoritative sources.

The alchemosexual symbolisms only grow stronger as the ritual proceeds, reinforcing the above interpretation, for now begins the component ritually recalling the murder of Hiram Abiff, and the ritualized "resurrection" of the new Master Mason. This part of the ritual begins by the gathered brethren feigning an end to the ceremony, and a pause for "refreshment:"

(Worshipful Master): - Brother Junior Warden, what is the hour?
(Junior Warden): - High twelve, Worshipful.
(Worshipful Master): - If you are satisfied it is high twelve, *you will erect your column*, and call the craft from labor to refreshment, for the space of thirty minutes (or fifteen minutes, as the case may be), calling them in at the sound of the gavel. On receiving this order, the Junior Warden takes from his desk a small wooden column, about eighteen inches in length, and sets it in an upright position at his right hand, and at the same time he gives three raps(***) with the gavel, and says:
(Junior Warden): - Brethren, you are accordingly at refreshment.
It should be remarked here, that there is a similar column on the Senior Warden's desk, which is always placed in a horizontal position(i.e., turned down on its side) when the Junior Warden's column is up, and *vice versa.* When the Lodge is opened, the Junior Warden's column is turned down, and the Senior Warden's turned up, at his right hand.[29]

It is difficult indeed to avoid the phallic, alchemosexual imagery in play here.

At this juncture follows the long ritual re-enactment of the murder of the Master Mason and architect of Solomon's Temple, Hiram Abiff, by the trinity of murdering masons - Jubela, Jubelo, and Jubelum - who try to force him to reveal the secrets of the third degree. When Hiram refuses, he is murdered. The new initiate, who has again been hoodwinked - blindfolded - plays the role of Hiram, and quite literally made to fall into a canvas, in which he is wrapped and then carried by the other brothers on their shoulders to his "grave."[30] He is then raised by the "Worshipful Master" after a prayer and a hymn, and the utterance of the "Lost Word of Masonry."[31]

For our purposes, it is worth noting the progression from the first to the third degrees, for in the first, the candidate is informed of the Compass, a symbol which in Pike's Scottish Rite becomes the symbol both of the primordial Geometer, God, but also of His primordial alchemosexuality. This is gradually unfolded in the second degree, then fully revealed - for one willing to "go beneath the surface"[32] - in the third degree, the degree of full resurrection and illumination. This suggests that the goal of this whole process of ritual resurrection has been to return man to a condition of "primordial alchemosexuality," prior to the division of the sexes. Nature, in this case, is the revelation of the original androgynous Geometer, the Grand Architect of the Universe.

B. The Rosicrucians

Such alchemosexual rituals are not unique to Freemasonry, however. They form, oddly, an almost universal component to all such fraternities, so much so, that one begins to wonder if the "disconcerting" and vaguely alchemosexuality of it all is accidental, or if there is a deeper connection. It was former Hoover Institute Fellow and scholar Anthony Sutton who first publicized yet another fraternal society, Skull and Bones,' preoccupation with rituals focused on death and homoeroticism. According to Sutton there are four elements of the initiation ceremony of Skull and Bones:

- that the initiate has to lie naked in a sarcophagus,
- that he is required to tell the "secrets" of his sex life to fellow initiates,
- that Patriarchs dressed as skeletons and acting as wild-eyed lunatics howl and screech at new initiates,
- that initiates are required to wrestle naked in a mud pile[33]

But there is another prominent secret society within western tradition, one in which the primordial androgyny and the descent of man from it, is explicitly taught: the Rosicrucians.

"Magus Incognito" was the pseudonym of William Walter Atkinson (1862-1932), an American attorney and occultist writing under a variety of pseudonyms.[34] In a short book, *The Secret Doctrine of the Rosicrucians*, "Magus Incognito" gives a brief, and indeed, classical review of basic esoteric doctrine. He begins his treatment by noting that in order to understand any esoteric doctrine, one must be prepared "to read between the lines of the text, and to reason by Analogy."[35] It is this process that allows an initiate to construct the laws of consciousness, and of the physical medium, the "laws of nature."[36]

But then comes a typical claim, a claim often made in occult and esoteric texts:

> The old Masters who made it the object of their lives *to gather together once more these scattered fragments, and to thus reconstruct the Occult Doctrine of the Atlanteans*, found a portion of their material in Egypt, in India, in Persia, in Chaldea, in Medea, in China, in Assyria, and in Ancient Greece, and also in the mystic records of the Hebrews, such as the Kaballah and the Zohar. The common source, however, may be regarded as distinctly Oriental. The great philosophies of the East, in fact, may be said to have been built upon the base of these still more ancient teachings. Moreover, the great Grecian Secret Teachings are believed to have been based upon knowledge obtained from this same common source. So, at the last, *the Secret Doctrine of the Rosicrucians may be said to be the Secret Doctrine of Atlantis, transmitted through the descendants of the people of that great centre of occult knowledge.*[37]

By referencing the "scattered knowledge" of the ancients, i.e., a "common core" of knowledge that had been the possession of "Atlantis," Magus Incognito is implying the reversal of the Tower of Babel moment, for by gathering the "scattered fragments" of this knowledge, the ultimate aim is to reconstitute the lost unity of mankind, a lost unity that as we shall see will involve all the familiar images encountered thus far. The second thing one must note is that, like the Freemasons, the Rosicrucians claim to have preserved elements of this doctrine within their fraternity *from before the Tower of Babel Moment.* It is important to note that Incognito *does not* claim that the Rosicrucians themselves have done this, but rather that the survivors of "Atlantis" have done this, and that this secret transmission of knowledge and doctrine ultimately issues in the secret doctrine of the Rosicrucian Fraternity.

Briefly put, Magus Incognito is maintaining that an elite was established after the "Tower of Babel Moment," after the fall of "Atlantis," and that this elite was tasked to preserve the core doctrines that made the advanced civilization of "Atlantis possible."

But what exactly *was* that doctrine?

Not surprisingly, the core of that doctrine is the by-now-familiar topological metaphor of the medium:

> In the Secret Doctrine of the Rosicrucians we find the following Aphorism of Creation:

The First Aphorism

I. The Eternal Parent was wrapped in the Sleep of the Cosmic Night. Light there was not: for the Flame of Spirit was not yet re-kindled. Time there was not: for Change had not re-begun. Things there were not: for Form had not re-presented itself. Action there was not: for there were no Things to act. The Pairs of Opposites there were not: for there were no Things to manifest Polarity. The Eternal Parent, causeless, indivisible, changeless, infinite, rested in unconscious, dreamless sleep. Other than the Eternal parent there was Naught, either Real or Apparent.

In this First Aphorism of Creation the Rosicrucian student is directed to apply his attention to the concept of the Infinite Source of All Things - the Eternal Parent - the Infinite Unmanifest, is represented by the Rosicrucians by the symbol of a circle, having nothing outside of itself and nothing within itself.[38]

In other words, like the mythological versions of the metaphor that we encountered in chapter two, the Rosicrucian version begins with a primordial Nothingness.

And like those mythological versions, the Rosicrucian version of this primordial Nothing contains within it "the possibility of infinite Thingess, or the infinite possibility of Things,"[39] an infinite potential. But then "Magus Incognito" also reveals, that for the Rosicrucians at least, there is a *physics basis* of the metaphor, a basis we encountered earlier in the *Hermetica*:

Infinite Space must be thought of as the Absolute Container of Everything, whether Manifest or Unmanifest - for outside of Infinite Space there is only Nothingess, or, more strictly speaking, there is no **outside** of Infinite Space.

Infinite Space, therefore, has always been the accepted occult and esoteric symbol by means of which men are able to "think of" the Infinite Unmanifest - the Eternal Parent, wrapped in the Sleep of the Cosmic Night. In one of the ancient occult catechisms, the question was asked : **"What is that which ever has been, is now, and ever shall be, whether there be a Universe or not, and whether there be gods or not?"** And the answer is: **"Space!"**

The strength of this symbol of Infinite Space, as indicating the Infinite Unmanifest, is perceived when the mind tries to think or even imagine, the absence of Infinite Space - either as absent before its creation, or else as absent after its destruction. it will, of course, be discovered that the human mind and the human imagination, finds it impossible to think of Space being absent in either event. The mind is compelled to think of Space as being Infinite and as being Eternal, without regard to whatever else is held to be either present or absent at any time, past, present, or future...

Moreover, as Infinite Space is invisible and beyond the other senses, it cannot be "known" or cognized as a Thing. Thought regarding it must always report "not this; not that" regarding it; and it answers to the ancient sage's statement of Reality that: "The Essence of Being is without attributes, formless, devoid of distinctions, and unconditioned. It is different from that which we know, and from that which we do not know..."[40]

Space, in other words, may only be known by means of the *via negativa*, the way of negations - defining not what it is, but what it is not - the way normally applied by mystics to the divine, to God. Rosicrucianism, in other words, has called the bluff by pointing out that the method is equally applicable to space, and that there is an underlying physics to the topological metaphor, for as was noted in the "First Aphorism," absolutely no distinctions or any distinct things apply to it.[41]

It comes as no surprise, therefore, that this Primordial Nothing, the Infinite Unmanifest, is also a primordial androgyny in its first manifestations:

This Bi-Sexual Universal Being, combining within itself the elements and principles of both Masculinity and Femininity, is known in the Rosicrucian Teachings as "The Universal Hermaphrodite," and "the Universal Androgyne."[42]

It is crucial to note what is occurring in the Rosicrucian version of the metaphor:

1) The primordial Nothingness is an androgyny;
2) The primordial Nothingness is also a metaphor for infinite Space
3) All differentiations arise from it.

In short, the alchemosexual metaphor is also a physics metaphor, and vice versa.

Moreover, within Rosicrucian symbolical lore, the circle and cross, and even the swastika, become symbols of this primordial androgyny.[43] But most importantly, this "bi-sexual" androgyny, or alchemosexuality, "operate and manifest upon every plane of Life, from the Sub-Minderal, on to the Mineral, on to the Plant, on to the Animal, on to the Human or to the Super-human, on to the Angelic or God-like."[44] Note the subtle implication here, for as mankind re-ascends the ladder of his descent, the inevitable goal is both an androgynous being, and an androgynous "consciousness," i.e., a consciousness that thinks in terms of triadic structures, of fusions:

> ...these Planes of Consciousness are known to the wise as (1) The Plane of the Elements; (2) The Plane of the Minerals; (3) The Plane of the Plants; (4) The Plane of the Animals; (5) The Plane of the Human; (6) The Plane of the Demi-Gods; (7) The Plane of the Gods.[45]

Once again, the alchemosexual symbol of androgyny becomes a symbol not only for fusions of all kinds, but an alchemical symbol in the proper sense as an image of the transformation of humanity, and its consciousness, themselves.

At this juncture, it is worth considering the principle symbolism of this transformation of consciousness in "Magus Incognito's" presentation, a symbolism with obvious Christian trinitarian roots:

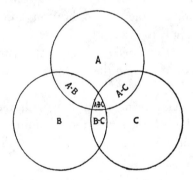

Magus Incognito's Trinitarian Consciousness Diagram[46]

He explains this symbolism as follows:

> Your attention is called to the fact that each circle in the symbol is called to and blended with the one on either side of it. Accordingly in the circular extent of each circle there is to be found FOUR different spaces or regions, as follows: (1) Its own unblended space or region; (2) the space or region in which its own space or region is blended with that of one of the neighboring circles, which constitutes a shield-shaped space; (3) the space or region in which its own space or region

is blended with that of the other neighboring circle, constituting a shield-shaped space; and (4) the space or region in the very centre of the symbol, in which the space or region of each circle is blended with that of both of the other two - thus producing a Triune Region. This arrangement, again, furnishes us with SEVEN distinct regions.... There are thus three unblended areas; also three blended areas of two elements; and finally one blended area of three elements.[47]

Note that we could also describe each blended region as a topological *surface* shared in common between two(or three) regions. In other words, Magus Incognito has perceived the deeply mathematical nature of the metaphor. But why draw upon a symbolism with obvious Christian roots?

C. Joachim of Fiore, the Hidden Androgynous God, and the Antinomian Kingdom of the Spirit

Let us return for a moment to literature: How does one rationalize Oscar Wilde's esoteric connections? How does one rationalize his making of experience of alchemosexuality a primary factor in his aesthetic philosophy? A clue, perhaps, is once again afforded in a comment of his Oxford mentor, literary and art critic Walter Pater, cited here at length to illustrate the strange context in which a passing reference emerges:

One of the strongest characteristics of that outbreak of the reason and the imagination, of that assertion of the liberty of the heart, in the middle age, which I have termed a medieval Renaissance, was its anti-nomianism, its spirit of rebellion and revolt against the moral and religious ideas of the time. In their search after the pleasures of the senses and the imagination, in their care for beauty, in their worship of the body, people were impelled beyond the bounds of the Christian ideal; and their love became sometimes a strange idolatry, a strange rival religion. It was the return of that ancient Venus, not dead, but only hidden for a time in the caves of the Venusberg, of those old pagan gods still going to and fro on the earth, under all sorts of disguises.... More and more, as we come to mark changes and distinctions of temper in what is often in one all-embracing confusion called the middle age, that rebellion, that sinister claim for liberty of heart and thought, comes to the surface. The Albigensian movement, connected so strangely with the history of Prevencal poetry, is deeply tinged with it.
.... It influences the thoughts of those obscure prophetical writers,

like Joachim of Flora, strange dreamers in a world of flowery rhetoric of that third and final dispensation of a "spirit of freedom," in which law shall have passed away.[48]

We have, as has been seen, argued that for Pater as for the other late nineteenth century critics and Uranians, "sentimentality" and "comraderie" are code words for the forbidden alchemosexual subject, and for the "antinomian revolt" against the Church and its morality that it implied.

But why, in this context, does Pater refer to the obscure medieval scholar of biblical prophecy, Joachim of Flora(ca. 1130-1200)? This is where the story gets interesting, laying bare the alchemosexual and hermetic imagery hidden within the Christian Trinity, and exposing the logic that would allow Pater to speak of a final dispensation of a "spirit of freedom" in which the Church's law and morality "shall have passed away."

Viewed a certain way, the essence of Joachim's system of "prophecy" is but the application of the "topological metaphor" to the historical process itself, viewing that process as a kind of "dialectical divination," which Joachim called "spiritual understanding." Joachim might, indeed, be called the first modern dispensationalist, for his writings are full of ornate - one is tempted to say, Baroque - "bible maps of the ages," anticipating by centuries the fundamentalist obsessions of modern American evangelicalism.

A glance at the pictogram used to depict the doctrine of the Trinity, in whose context Joachim conceptualized these ideas, will be helpful to show its deeply alchemosexual basis, and why, in turn, that basis would have been seen by Walter Pater as a code for the alchemosexual *experience* and the antinomian spirit of freedom.

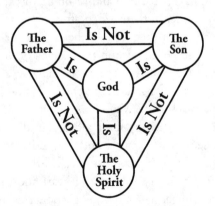

The Common Western Pictogram of the Holy Trinity

Hidden within this pictogram is a careful topological metaphor of the hidden androgyny, an alchemosexuality that, like all the others examined previously, creates information from within itself, for by Joachim's time, it was standard western trinitarian doctrine that the Son took his origin from the Father, and the Holy Spirit took His origin, in turn, from both the Father and the Son, becoming, in St. Augustine's words, the "consubstantial love of both,"[49] or, in the terms of our topological metaphor, the common surface between the two. In other words, we have once again a primordial "nothing" that self-differentiates into a one-three of two regions of distinguished nothing - Father and Son - sharing a common surface, the Spirit:

1) The first region of differentiated Nothing, The Father, $\varnothing^\circ_F$;
2) The second region of differentiated Nothing, The Son, $\varnothing^\circ_S$; and,
3) The common *surface* of differentiated Nothing shared between the two, the "consubstantial love of both," to cite Augustine once again, the Spirit, $\partial\varnothing_{Sp}$.

But is this really a new instance of the analogical process of the topological metaphor that we discussed previously? We believe that it is, and for one very important reason. Let us recall the metaphor as it occurred in the previous texts examined. There we pointed out that each of the "differentiated Nothings" shared *common functions* that made them all analogies of the other in the following passage from the *Hermetica:*

> Of what magnitude must be that space in which the Kosmos is moved? And of what nature? Must not that Space be far greater, that it may be able to contain the continuous motion of the Kosmos, and that the thing moved may not be cramped for want of room, and cease to move? – *Ascl.* Great indeed must be that Space, Trismegistus. – *Herm.* And of what nature must it be Aslcepius? Must it not be of opposite nature to Kosmos? And of opposite nature to the body is the incorporeal…. Space is an object of thought, but not in the same sense that God is, for God is an object of thought primarily to Himself, but Space is an object of thought to us, not to itself.[50]

Now let us consider Augustine's version of this commonality of functions in his derivation of the Holy Spirit from the Father and the Son, for once again, the analogical nature of the topological metaphor of the medium is reproduced exactly. Augustine states:

> For we cannot say that the Holy Spirit is not life, while the Father is life, and the Son is life: and hence as the Father ... has life in Himself; so He has given to Him (the Son) that life should proceed from Him, as it also proceeds from Himself.[51]

Here the common function shared by the Father, Son, and Spirit, is obviously "life in Himself," or asiety in the Latin, The *differentiating* function is that the Spirit proceeds from two classes of causes, an Uncaused Cause, the Father, and a Caused Cause, the Son. Small wonder then, that the trinitarian pictogram reproduced by the medieval western church so closely resembles the structure in the *Hermetica,* for the two structures are, in fact, one and the same, and emerge from the topological metaphor itself and from the analogical method that it implies; it does not emerge from "revelation."

It is in this rich metaphorical matrix that Joachim's thought emerges, and with it, why Pater would have so carefully chosen his reference to it as a code for the alchemosexual ideal and its fraternal continuity through history. Basically put, Joachim's conception is quite simple:

> In his speculation the history of mankind had three periods corresponding to the three persons of the Trinity. The first period of the world was the age of the Father; with the appearance of Christ began the age of the Son. But the age of the Son will not be the last one; it will be followed by a third age of the Spirit.[52]

In Joachim's view, in that third age of the Spirit, the socio-ecclesiastical types of laymen and cleric would be blended into the emergence of a new type of social order, wherein "all religious orders" would perish, leaving only a monks - the "androgynous" synthesis of laity and clerics, just as the Spirit issues from the Father and the Son - to survive.[53]

The imagery here is deliberately, though subtly alchemosexual, as Pater almost alone and uniquely of all scholars of the Middle Ages perceived, for Joachim does not prognosticate on *how* such a society of monks would reproduce, but by subtly implying the alchemosexual image, he has revivified and resurrected the ancient cosmologies of the primordial alchemosexual, masculine-androgyny. In his hands, this new spiritual man would abolish classes of authorities - bishops, popes, princes, and laity - and substitute it with a community of immediate individual experience of communion with the divine, a "fraternity" of "the spiritually perfect who can live together without institutional authority."[54] This "antinomianism," to use Pater's words, is thus another code for what is really taking place: the alchemical transformation of man *back* into the primordial alchemosexual being from which - according to the metaphor - he originally emerged as a masculine androgynous creature.

Consequently, by his deliberate reference to the Albigensians, against whom the charge of alchemomosexual practices was leveled,[55] and more importantly, by deliberate reference to Joachim, where "Spirit" represented the final antinomian freedom of the androgynous synthesis proceeding from the Father and the Son, Pater is really implying that the idealized feminine loves within this poetry - for example, of Dante's Beatrice - is a *code* for the alchemosexual comraderie and experience of "the few," the initiates into a hidden brotherhood of alchemosexuality. Joachim, in his deduction that the metaphor will issue in a coming age of Spirit, in which there is no need for priesthoods or churches, has deduced the obvious, and one might say, the inevitable conclusion from the premise of the metaphor, for in a world that emerges as a process of differentiation of the primordial alchemosexual Nothing, no one is in a privileged position to represent Him, nor are sacrifices, or priesthoods, required to approach Him, for everyone literally interfaces directly with Him. In this, he has looked back to the Platonic and Greek ideal alluded to by Pater, and in this, he has also anticipated, by several centuries, the doctrines of Illuminism and Freemasonry, as we shall now discover. And it should not surprise us, now, that Aquinas was one of the few mediaeval schoolmen who did not condemn Joachim's "apocalyptic androgyny" as being heresy.

While the alchemosexual rites and doctrines surveyed in this chapter may seem coincidental or at best synchronous, there is a much deeper connection between the Lodge and the nineteenth century literary movement known as the Uranians, as we shall discover in the next chapter, a connection that once again, unbelievably, takes us back to the University of Oxford, and Oscar Wilde, whose celebrated novel of genius, *The Picture of Dorian Gray*, reveals a true encoded alchemosexual program, a veritable *fin de siecle* eschatology.

To state the question succinctly: Why should the nineteenth century have seen such an outburst of these ideas, from Pater's careful and subtle suggestions in his literary criticism, to Percy Bysshe Shelley's *Frankenstein*, to Wilde's *Picture of Dorian Gray*?

The answer is as astonishing as it is disconcerting, for it once again returns us to biology, to genetics, and to physics, and to the possibility that the original alchemosexual androgynous metaphor was born in a sophisticated scientific culture.

ENDNOTES

1   Walter Pater, *The Renaissance: Studies in Art and Poetry* (Dover Publications, 2005), p. 21.

2   Citing Robert Boyle, *Works*, Robert Lomas, *Freemasonry and the Birth of Modern Science* (Fair Winds Press, 2003), p. 65

3   Malcolm C. Duncan, *Duncan's Masonic Ritual and Monitor* (Kessinger Legacy Reprints, No Date, ISBN 9781162563527), p. 28.

4   Ibid., p. 30.

5   Ibid., pp. 34-35, emphasis in the original. The purpose of Duncan's emphasis was to draw a connection to the Pythagorean Hipparchus, who betrayed the Pythagorean brotherhood, and committed suicide, his body lying on the sandy beaches of a seashore, implying a connection and continuity of Masonry with the Pythagoreans. See p. 35, note 2.

6   Ibid., p. 35.

7   Albert Pike, *Morals and Dogma*, p. 653.

8   Ibid., p.

9   Ibid., p. 653.

10   Ibid.,p. 851, emphasis added.

11   For a brief discussion of the social engineering aspect of the symbolism of the rough and perfect Ashlars within Masonry, see Joseph's *LBJ and the Conspiracy to Kill Kennedy* (Adventures Unlimited Press, 2011), pp. 220-224.

12   Malcolm C. Duncan, op. cit., p. 59.

13   Ibid., pp. 60-61.

14   Ibid.,pp. 63, 58.

15   Ibid., p. 65.

16   Ibid., pp. 72-73.

17   Ibid., p. 74.

18   Ibid.

19   Ibid., pp. 77-78, emphases added.

20   Ibid., p. 79, emphasis added.

21   Ibid., p. 85, emphasis added.

22   Ibid., p. 87.

23   Ibid., p. 88, emphasis added.

24   Ibid., p. 87.

25   Ibid., p. 89.

26   Ibid., p. 90.

27   Ibid., p. 96.

28   Ibid., p. 99, emphasis added.

29   Ibid., p. 100, emphasis added.

30   Malcolm C. Duncan, op. cit., pp. 101-120.

31   Ibid., pp. 119-121.

32   Oscar Wilde, "Preface" to *The Picture of Dorian Gray* (New York: Barnes and Noble, 2003), p. 2.

33   Anthony Sutton, *America's Secret Establishment: An Introduction to the Order of Skull and Bones* (Billings, Montana: Liberty House Press, 1986 ISBN 0-937765-02-3), p. 201.

34   "William Walter Atkinson," *Wikipedia*, http://en.wikipedia.org/wiki/William_Walker_Atkinson

35   Magus Incognito(William Walter Atkinson), *The Secret Doctrine of the Rosicrucians* (Cosimo Classics, 2010), p. 13.

36   Ibid., p. 15.

37   Ibid.,p. 17, emphasis added.

38 Ibid., p. 23, emphasis in the original.

39 Ibid., p. 25.

40 Ibid., pp. 25-26.

41 Ibid., p. 30.

42 Ibid., p. 52.

43 Ibid., pp. 55-56.

44 Ibid., p. 56.

45 Ibid., p. 85.

46 Ibid., p. 118.

47 Ibid., p. 143.

48 Walter Pater, op. cit., pp. 21-22, emphasis added.

49 St. Augustine, *On The Trinity*,

50 *Libellus: 1-6b, Hermetica,* trans. Walter Scott, Vol. 1, pp. 135, 137.

51 St. Augustine, *On the Trinity*, 15: 27:48, in *The Nicene and Post-Nicene Fathers*, ed. Philip Schaff.

52 Eric Voegelin, *The New Science of Politics: An Introduction* (Chicago, 1987), p. 111.

53 Marjorie Reeves, *The Influence of Prophecy in the Middle Ages: A Study in Joachimism* (Nore Dame: the University of Notre Dame Press: 1993), p. 203.

54 Voegelin, op. cit., pp. 112-113.

55 The accusation of alchemosexual practices was leveled at the Albigensians (Cathars) by the medieval western Church, an accusation that surfaced again against the Templars after King Philippe le Bel closed the order down.

❧ Eleven ❧

THE ANDROGYNOUS ALCHEMOSEXUAL APOCALYPSE:

HERMAPHRODITISM, URANIANS, SHAMANS, AND GENETICISTS

⁛

*"Man is a bisexual organism that forgot its bisexual origin. The problem with which the alchemist - as well as Surrealism and analytical psychology are concerned - is to bring this shattering reality back to man's consciousness; in other words, to endow man with the **awareness**, the alchemical **aurea apprehensio**(golden awareness) of this marvellous reality: we are gods, because we all are man and woman at one and the same time.*

*"In all mythologies, gods are immortal and androgynous. As a matter of fact, gods are immortal **because** they are androgynous."*
—Arturo Schwarz[1]

SUCH CLAIMS AS ARE ADVANCED within esoteric doctrine for androgyny and immortality, as evidenced in the epigraph above, seem at first glance both fanciful and downright incredible, and yet, the further we delved into the topic, the more astounded - and incredulous - we became. As will be outlined in this chapter, we discovered government agencies investigating homosexual neurophysiology and genetics; whole literary movements doing the same; an explosion of actual hermaphroditism in the nineteenth century and an ensuing medical-taxonomical controversy among mystified physicians; shamans (and anthropologists) talking about androgyny; and even a kind of genetic, "embryonic" androgyny. And of course, we encountered a whole host of articles from religious circles opposed to all of it in the name of implied Aristotelian metaphysical principles, and the Bible.

We were literally stunned by all this, but as we pressed our research, the shock, and the accompanying questions, only grew. For example, the U.S. Army, we learned, was researching methods on how to regrow limbs and to repair burns. That much, we knew already, was standard fare. But then we read that one of the methods being researched

> uses immortal cells from a newborn's foreskin. Unlike normal cells, which have a natural limit on the amount of times they can divide and duplicate, immortal cells can keep duplicating forever unless killed by an outside force. NEKs(normal keratinocytes) cells that are found in the circumcised foreskin of babies, have been found to be immortal and, unlike most donor skin, can be placed on a wound without being rejected by the recipient's body.[2]

Foreskin? Immortality cells? Was there some deep, ancient connection between this fact and the rite of circumcision practiced in so many religions, rites that in some cases hinted at a connection between the act of circumcision and the immortality of God?

The esoteric tradition, as noted in the epigraph that began this chapter (not to mention the Mayan *Popol Vuh* cited in chapter one), indicated that mankind's original state was as an androgynous creature, and this in turn was linked to his immortality. And this androgyny, we quickly discovered, was apparently one goal of the scientists in their laboratories. For example, a BBC news article from 2003 reported that scientists in the USA were successful in creating "a mixed-sex human embryo," an androgyne. The reason offered for this bizarre experiment was to discover if they could prevent "certain genetic diseases from arising." And then, having created these androgynies, we are told that "the 'merged' embryos were never intended to develop into children, and were destroyed after a few days."[3] Note the moral and ethical implication being reached here, without so much as a discussion: androgynous humans are somehow less than human, a conclusion, as we shall shortly discover, is mirrored in the early nineteenth century medical discussions of hermaphroditism.

Further investigation revealed that a California company was using "cloning technology to make five human embryos" for the purpose of "harvesting stem cells." The embryos were created by using skin cells of two men who worked at the company, and the company "painstakingly verified that the embryos were clones of the two men." Having apparently decided that these embryos were less than human, in spite of being clones of their donors, the company eventually destroyed the embryos.[4] Note once again how the modern example mirrors the mediaeval and early Renaissance discussions: was someone born "virginally" and without normal human sexual intercourse fully human? And the answer given by this modern corporation's practice would seem to be "no."

A review of all the data began to unfold a disquieting possibility: could the ancient metaphor, with all its emphasis on the primordial androgyny both of God and of man, actually have been formed by a scientifically advanced culture, on the basis of applying analogical thought to the principles of biology and physics? Was it, in other words, a metaphor for principles not only of a "topological" metaphor of the medium, but for those of biology as well? And if so, was there a connection between the two? For whatever else might be said, it was clear that scientists were engaged in nothing less than the creation of not only of the transhumanist cyborg, but also of the alchemical "masculine androgynous homunculus" born "virginally" of a scientific technique.

This was brought home with especial force after a visit to the website of the Oakridge National Laboratory, where we discovered a link to the Behavioral Genetics component of the Human Genome Project.[5] There, we read the following, with the underlined portions referring to linked papers in behavioral genetics:

> Online Mendelian Inheritance in Man (OMIM) is a large, searchable, up-to-date database of human genes, genetic traits, and disorders. Each OMIM record contains bibliographic references and a summary of the scientific literature describing what is known about a particular gene, trait, or disorder. The following behavioral traits are included in OMIM. The six-digit number MIM number is used to uniquely identify each record.
>
> - Hand skill, relative (handedness): (139900)
> - Hand clasping pattern: (139800)
> - Arm folding preference: (107850)
> - Ears, ability to move: (129100)
> - Tongue curling, folding, or rolling: (189300)
> - Musical perfect pitch: (159300)
> - Novelty seeking personality trait: (601696)
> - Stuttering: (184450)
> - Tobacco addiction: (188890)
> - Alcoholism: (103780)
> - Homosexuality: (306995)[6]

An odd list, to say the least, since its implications are that the ability to curl one's tongue, or to have perfect musical pitch, or to be addicted to alcohol, tobacco, or to stutter, or to be a homosexual, were all being studied by geneticists.

Why study homosexuality in the context of behavioral genetics at all? In seeking an answer to that question, we discovered a long, winding, trail of research, and a stunning conclusion. It began with a little-known fact of medical history, what we began to call:

A. The Nineteenth Century Explosion of Hermaphroditism

And it *was* an explosion, a drastic increase in the numbers of people whose sex, for whatever reason, was not easily determinable, and it provoked another explosion: a controversy over how to *classify* such people medically. Were they male? female? or, as some dimly began to suspect and quietly voice, something in between: androgynes. The question was more than just a question of medical taxonomy, for one's status in society, and in *law*, was to some extent dependent upon the answer to that question.

To put the point as simply as possible, the doctors of the nineteenth century began to treat patients whose sexual characteristics seemed to be disquietingly androgynous, and these human curiosities oftentimes volunteered to have their bizarre genitalia photographed for medical journals. Nor was the phenomenon localized; it began to be reported in France, Germany, Poland, and England. Alice Domurat Dreger, in a seminally important study, *Hermaphrodites and the Medical Invention of Sex*, published by the Harvard University Press, summed up the explosion, and consternation of physicians trying to explain the phenomenon, as follows:

> In the last few years of the nineteenth century, there occurred a virtual explosion of human hermaphroditism. Why did this happen? Some people have suggested to me that a significant increase in industrial pollution could have contributed to the apparent rise in numbers of cases of hermaphroditism, but it is very hard to know what, if any, significant material environmental changes might have given impetus to the rise.... Instead I think it reasonable to credit the steady rise to other sorts of important social changes.[7]

The occurrence of the phenomenon had medical professionals progressively refining their criteria for sexual classification, an enterprise that grew increasingly uncertain, for "Indeed, the more confusing and abundant cases of doubtful sex became, the more certain and constrictive concepts of true hermaphroditism grew.... Medical men triumphantly reconstructed and constricted the true hermaphrodite even while - perhaps because - they witnessed and were forced to confess serious doubts about sex."[8] In other words, the more medical taxonomy tried to force individuals into the "natural" categories of male and female, the more the phenomenon of hermaphroditism actually grew, as ambiguous cases that fell between the cracks of the criteria increased by dint of the criteria themselves.[9]

So how does one explain the apparent sudden rise of the phenomenon? Dreger maintains that one reason is simply due to the increase of availability to medical care and particularly to the emerging medical field of gynecology. Inevitably, this meant an increase in the number of reports of hermaphroditism.[10] Dreger also observes that there were increasing means for *reporting* medical anomalies; as medical publications increased, so did reports of androgyny. As a consequence of this, the conviction grew amongst medical professionals that hermaphroditism was actually "not all that rare."[11]

Thus far, then, we have the following explanations for The Nineteenth Century Explosion:

1) The rise of hermaphroditism was more apparent and accidental, than due to more material underlying causes, since it resulted from the expansion of the availability of medical care and the media for reporting anomalous cases; or,
2) It was the result of evolutionary responses to the environment (in the form of increasing industrial pollution).

However, as is by now evident, the metaphor of androgyny is very old, and there are suggestive if not conclusive indicators that it might be the residue and legacy of an ancient scientific sophistication. This factor raises the possibility of other explanations for the phenomenon, and for its sudden rise in the nineteenth century:

3) The ancient classical societies discovered similar cases of androgyny, and extrapolated the androgyny metaphor from them; or,
4) The phenomenon of androgyny itself is a residue of mankind's actual primordial state[12] - as is actually claimed by various traditions - and occassionally manifests itself for whatever reason, including the possibility that it is being deliberately, though covertly, sought and engineered.

So which of these possibilities is the likely one?

It is in answer to this question that physicians confronted with The Nineteenth Century Explosion began to grope toward a disturbing resolution, for the phenomenon was a challenge to "medical and scientific concepts of the male and the female as well."[13] To put it more succinctly, what was at stake was the broad philosophical culture of Europe, influenced as its social mores and jurisprudence were by Christianity. Indeed, Dreger observes that the attempts to pigeon-hole hermaphrodites into either sexual category was being done predominantly by heterosexual male doctors,[14] and solely on the basis of "gonadism," rather than on the basis of the social experience of such individuals.

1. "Gonadism"

Not surprisingly, doctors tried to define the problem out of existence by what Dreger calls "Gonadism," i.e., the assignment to hermaphroditic humans displaying androgynous characteristics to either the male or female sex, based upon the proponderence of male or female sexual characteristics of their genitalia.[15] This definition was, for the nineteenth century, particularly helpful in cases where an individual's genitalia had all the appearance of a female vulva(in some cases complete with the apparent orifice) but, upon examination, were discovered to be but partially descended male testicles.[16] But such definitions were ultimately no help, for such individuals, by dint of their ambiguity, were often raised by their parents in the *opposite* social-sexual role than that assigned to them by the sexual determinations of "gonadism."

"Gonadism", however, quickly proved to have its other inherent limitations, for ambiguities might be, and were quickly discovered to be, difficult, since testes "might remain undescended, and ovaries might be found in unexpected places, too,"[17] and the Polish gynecologist Neugebauer pointed out precisely such cases of hermaphroditism where "an ectopic ovary has often been taken for a testicle, or a testicle delayed in its descent for an ovary."[18] The bottom line was that gonadism quickly collapsed as a criterion and strategy of taxonomy, for "Navigation of this sea of doubt was no job for the hasty. The sexes could look and feel remarkably similar even to the experienced medical man."[19] The admission is all the more remarkable in Neugebauer's case, for this Polish gynecologist was one of the promoters of the idea of "gonadism" and "pseudo-hermaphroditism" in the first place, i.e., of the idea that they possessed a "true sex" - male or female - masked by ambiguous "malformations."[20] Yet, at the same time, it was this same Neugebauer who asked the British Gynaecological Society to publish a 1903 paper of his called "Hermaphrodism in the Daily Pratice of Medicine: Being Information upon Hermaphrodism Indispensible to the Practitioner." As Dreger observes, the title alone indicates that "society felt that hermaphroditism was a common enough problem that nearly all practitioners engaged in 'the daily practice of medicine' needed an education in hermaphroditism."[21]

2. Androgyny: Pathology? or Third Sex?

Needless to say, the androgynous ambiguities that 19th century medicine began to explore, and the rise of "gonadism" as a taxonomical strategy, carried with it the implication that hermaphrodites were "monstrosities," a pathological departure from the "norm" of male and female.[22] This in part reflects the incipient "Aristotelianism" of nineteenth century medicine, for which a "nature" was a more or less fixed and permanent, and therefore more or less easily definable. phenomenon, notwithstanding the evolutionary assaults on that idea already underway. This led nineteenth century medical science to view females as "underdeveloped males", another legacy of Aristotle.[23] As we shall see in a moment, modern genetics has stood both notions on their heads.

a. "Embryonic Androgyny"

Some physicians, however, began to suspect that the hermaphroditism they were witnessing was something much more fundamental to human nature, for they already knew that in the first weeks of pregnancy, a human foetus displayed precisely such androgynous characteristics:

> ...they knew that ultimately-male and ultimately-female fetuses began with Mullerian and Wolffian systems of proto-organs internally. In the female, however, the Wolffian system atrophied and the Mullerian system evolved to form "female" internal organs, including the fallopian tubes, uterus, and vagina. In the male, the Mullerian system atrophied and the Wolffian system evolved to form "male" internal organs, including the deferent canals and the prostate. Knowledge of such common developmental pathways made it possible to explain, for instance, how a "true male" could seem to have developed an otherwise inexplicable vagina or uterus.[24]

To put it bluntly and succinctly, as The Nineteenth Century Explosion was occurring, medical science already knew that there was at least *some* truth to the ancient androgyny metaphors, in that *everyone, without exception,* began in as a kind of "embryonic androgyny."

Consequently, a few cautious voices within medicine began to suspect that androgyny was not a pathology, but perhaps something more fundamental:

> Hermaphroditic humans were also used as living proof of "the primitive bisexuality and the primordial hermaphroditism of the embryo. Even while the female was often portrayed rhetorically as a sort of underevolved male, medical and scientific men thus professed the conviction that the male and female types actually diverged from a common, original hermaphroditic state in both embryology and evolutionary history. The British surgeon Jonathan Hutchinson, for instance, suggested that "like many other conditions, hermaphroditism is a thing of degree, and .... up to a certain point all persons are bisexual." He reminded his readers "Up to a certain age the foetus has potentially the organs of both sexes, and it is only by the ascendant development of the one set that the other is suppressed. Nor is the suppresal(sic) in either sex ever absolute, for every male has mammary glands and every female has a clitoris, organs which definitely belong to the other sex, and which persist only because the suppressal has been imcomplete.[25]

These facts led the 1860s-1870s legal reformer Karl Ulrichs to suggest that every embryo that eventually manifested as a human hermaphrodite to have come about from a *lack* of development that occurred in "normal" people. It was, in other words, entirely *natural*, but not *normal*. Following up this position, Ulrichs suggestsed that every embryo contains "germs" for both types of sex drive, and that these, too, could persist in androgynes, leading to what was then called Uranianism, or what we would now call male homosexuality.[26] Ulrichs was one of the nineteenth century's many "Uranian" apologists - homosexual men studying the phenomenon, and trying to reason for a more tolerant attitude within society for such persons. He and his colleague, Xavier Mayne, whom we encountered long ago in the introduction to this book, were among the first intellectuals to draw the inevitable conclusion from The Nineteenth Century Explosion and the resulting medical controversy: it was the Aristotelian-Judeo-Christian definition of human nature itself that was being called into question.[27] This approach would be echoed again in the twentieth century by André Gide in his *Corydon.* Notwithstanding these efforts, the bulk of medical taxonomy continued to view hermaphroditism, and all extensions of the idea to male and female homosexuality, as a "decline" and even as an actual "falling away from the genus" *homo.*[28]

Nonetheless, slowly, albeit very quietly, the idea was growing that hermaphrodites and, among some Uranians, male and female homosexuals, were manifestations of the primordial "third sex", the true androgynes, an "intersex", for in the late twentieth century, as genetic confirmations of hermaphrodites carrying both XX female sexual cellular characteristics, *and* XY sexual cellular characteristics, began to be documented,[29] so did the suspicion that their occurrence may be quite natural, but simply not normal to most humans. Indeed, it could be argued, following this line of reasoning, on a *genetic* basis, that *all males*, because they carry the sexual determinants of both females and males in their testes, are to that extent, somewhat androgynous. Thus, we now have, in addition to the "embryonic" androgyny, yet another to contend with: a *genetic* androgyny.

But before we can confront modern genetics and what it has to say about all of this, we must deal with the other great Nineteenth Century Explosion, the Uranians, themselves.

B. The Uranian-Darwinian Guess
1. Edward Carpenter, Uranianism, and the Topological Metaphor

The Uranians were, according to some modern gay rights activists, precursors of their movement, and indeed, to a certain extent, that is true. But there all resemblance really ends, for unlike the modern movement, the nineteenth and early twentieth century Uranians were almost without exception men of learning and letters, and less interested in their "rights" than in trying to understand the possibly deep and ancient roots of their own "condition". They were more social philosophers interested in a particular human phenomenon and its implications, than they were a "movement" with a political agenda, though they were certainly aware of the political and social implications of their research and studies. Unlike writers such as Oscar Wilde, however, they also chose not to disguise their alchemical concerns in clever fiction or literary criticism, but to share them openly, in reasoned non-fiction.

Men like John Addington Symonds wrote openly about Platonic philosophy and the role of homosexual love, and more importantly for our purposes, the metaphor of androgyny within it, as a basis to challenge the persecution of homosexuals. It takes only a glance at the title of Symonds' chief work - *A Problem of Greek Ethics: Being an Inquiry into the Phenomenon of Sexual inversion, Addressed Especially to Medical Psychologists and Jurists* - to see that with The Nineteenth Century Explosion, a social revolution had begun in earnest, for when it was published in 1908, the famous trial of Oscar

Wilde was still a living memory. But the leading luminary of this movement was the philosopher Edward Carpenter (1844-1929).

Edward Carpenter, 1844-1929

Embracing socialist views, Carpenter also viewed "inverted love" as an equalizing phenomenon in class-conscious Britain. But if this were all there was to Carpenter, he would be dismissible as a minor prophet of a socialism and its implications.

But there was a much deeper philosophical component to Carpenter and to his advocacy of "Uranian" acceptance. One can, when reading Carpenter, find astonishing citations, often back to back, of authors one normally wouldn't connect, such as the Neoplatonic philosopher Plotinus, and the American poet Walt Whitman.[30]

In his controversial book, *The Intermediate Sex: A Study of Transitional Types of Men and Women*, Carpenter cites, approvingly, the observation of the French Uranian De Joux:

"We form," he says, "a peculiar aristocracy of modest spirits, of good and refined habit, and in many masculine circles are the representatives of the higher mental and artistic element.. In us dreamers and enthusiasts lies the continual counterpoise to the sheer masculine portion of society - inclining, as it always does, to mere restless greed of gain and material sensual pleasures."[31]

This is one of the first recurrences in modern times of an ancient theme, found in Plato's *Symposium*, that the "Uranian" individual's love was a "higher love," it was, indeed, as the name the Uranians took to describe themselves, "heavenly," from the Greek word for heaven, *ouranos,* from which the word Uranian derived.

It was, argued Carpenter, not a "choice," but something inherent to the individual experiencing it:

> ...the feeling is, as said, so deeply rooted and twined with the mental and emotional life that the person concerned has difficulty in imagining himself affected otherwise than he is; and to him at least his love appears healthy and natural, and indeed a necessary part of his individuality.[32]

In other words, for Carpenter - giving voice to the sentiments and conclusions of many 19th century Uranians - it was *not* a matter to be reduced to "gonadism," but was something much deeper, for it was *indicative of mental and emotional states unique to Uranians.* As we shall see shortly, there is some modern genetics confirmation of this view that, indeed, "Uranians" are neuro-physiologically unique.

It was this mental and emotional difference that Carpenter singled out in Plato as being the cause for this claim of being a "higher love:"

> Finally it seems to have been Plato's favorite doctrine that the relation if properly conducted led up to the disclosure of true philosophy in the mind, to the divine vision or mania, and to the remembrance or rekindling within the soul of all the forms of celestial beauty. He speaks of this kind of love as causing a "generation in the beautiful" within the souls of the lovers.[33]

Clearly Carpenter has moved beyond the typical Uranian call for a reassessment of social attitudes toward "uranianism," to something else, and quite different, from it: to the assertion that it is of a different and indeed higher quality.

The question is, why?

And the answer, once again, lies with the "topological metaphor," for Carpenter clearly singles it out, and in doing so, comes to one of the first modern articulations of a "deep physics" connecting Mind and Matter, or Mind and the Physical Medium itself. He begins his study of this topic, *The Art of Creation*, with some intriguing observations:

> We may say here, however, that the distinction between Mind and Matter forces us to conceive, or try to conceive, of a 'stuff' prior to both - a something of which they are the two aspects; and thus we come to the world-old idea of primitive Being (*before all differentiation, emanation, or expression*), or the 'Will' of the later philosophers (Schopenhauer, Hartmann, Royce, and others). This Will or Being is absolutely not thinkable by the ordinary consciousness (except as a necessary ground for other thoughts), for obviously it lies beyond the region of thought. I shall, however, endeavour to show that it *is* known in the stage of (cosmic) consciousness transcending our ordinary consciousness. The perception of matter and mind as distinct things belongs only to our ordinary (self_ consciousness. This distinction is not known in the earlier stage of simple consciousness, and it passes away again in the higher and more perfect stage of the cosmic consciousness.[34]

Thus, there was a third, and higher stage of consciousness, in which the distinctions of Mind and Matter were somehow overcome, and this, once again, speaks to the power of the symbol of androgyny as a metaphor for the fusions of all types of distinctions.

But from this now-familiar metaphor, Carpenter draws a social implication, one inimical to the later Yahwist traditions which overturned it, and in so doing, exposes what may have been the real reason for the Uranian persecutions:

> Here, in the contemplation of this universal Being, this primal Self of all, we are at the source of Creation. In this primal Self, and its first differentiation, we may suppose to exist great primitive Ideas, attitudes, aspects - things below or more fundamental than Feeling, Thought, and Action. These ideas are working everywhere - in the great Self, **and in every lesser self that springs therefrom**; and our lives *are* their expression (differently mingled though they be in each person, and always, owing to the conflict of existence, *inadequately* expressed).[35]

Carpenter has exposed the implication of the ancient topological metaphor, a metaphor which we have also discovered in previous pages is the androgyny metaphor, and that implication is that if everyone, by dint of their "descent" from the physical medium, is a direct manifestation of it, then there is quite simply no need for special revelations, religions, Scriptures or "sacred texts;" the text is nature, and the God is Nature's God. Thus, Carpenter is also exposing yet another implication, for it is equally true that no *fraternal or secret society tradition* is a unique program or revelation either. The apocalyptic bloodlettings of the French Revolution, with its own appeals to deep esoteric traditions, are equally indicted.

This point may not be readily apparent, however, without a consideration of Carpenter's exposition of Consciousness and its relationship to the medium. As was seen in chapter two, the initial differentiation results always in a basic triadic structure, which, given the connection between Mind and Medium, is also according to Carpenter a basic triadic structure of consciousness:

> Every act of knowing involves three aspects, which we cannot avoid, and under which (by the present nature of our minds) we are forced to regard it. There is (1) the knower or perceiver, (2) the knowledge or perception, (3) the thing to be known or perceived. I say we cannot imagine the act of knowledge or perception except in this triple form.[36]

To make this point clear in terms of the quasi-formal notation with which we have expressed the metaphor, it would look like this:

1) The Knower and the object or thing to be known, are the two interior regions: $\varnothing_x$ and $_y$ respectively; and,
2) the knowledge itself, represented as the common surface linking the two: $\partial\varnothing_{x,y}$.

There is, moreover, as Carpenter also observes, a "root" Consciousness[37] or "underlying Ego" or Self,[38] symbolized as the underlying $\varnothing$ undergoing differentiation, such that creation itself is a perpetual process of differentiation and perpetual consciousness.[39]

In the first stage of consciousness, Carpenter maintains that the state is that of a "Simple Consciousness" (note the Neoplatonic term), in which "the knower, the knowledge, and the thing known are still undifferentiated,"[40] a stage that Carpenter believes that animals possess.[41]

The second stage has, as it were, two "sub-stages," best exemplified in human psychological development:

The second stage is that in which the great mass of humanity at present is; it is that in which the differentiation of knower, knowledge, and thing known has fairly set in.

... Its arrival can generally be noticed without difficulty in any young child. It is the beginning of a new era in its development, and from that moment life begins to shape round the self.

But at the same moment, or very shortly after, the child begins to recognise the self in others - in its mother and those around. And - what is curious and interesting - the child ascribes 'selves' also to toys, stones, and what we call inanimate things. In fact, simultaneously with the appearance of the subject in consciousness comes the appearance of the object in consciousness. It is curious that at these early stages the object of knowledge and the knowledge should be differentiated from each other, or begin to be differentiated; but it is so. The child feels not only (as we do) that there is a personality behind the appearance of its mother, but that there is something behind these stocks and stones, and personifies them also. So does the savage.[42]

The second part of this stage, at least in Western culture, is reached later when "consciousness" is removed from inanimate objects.

It is at the third stage, that Carpenter exposes what we have noted all along about the metaphor: it is comprised of a both/and, not an either/or dialectic, and it happens in a kind of mystical synaesthesia, a reintegration.[43] But this is decidedly *not* reabsorption and a loss of the distinctive self, but rather, a reintegration based - as it was in Shelley and Wilde - in a kind of communal love, for love can exist only where there is differentiation, and yet, also only exists when there is reintegration:[44]

> "Knowledge has three degrees," says Plotinus — "opinion, science, illumination... It (the last) is absolute knowledge founded on the *identity of the mind knowing with the object known.*"
>
> "God is the soul of all things," says Eckhardt, "He is the light that shines in us when the veil (of division) is rent."
>
> Whitman speaks of the light that came to him: -
> "Light rare, untellable, lighting the very light,
> Beyond all signs, descriptions, languages,"
> And says -
> "Strange and true, that paradox hard I give,
> Objects gross and the unseen soul are one."[45]

These insights lead Carpenter to some very direct and provocative statements on the relationship of consciousness to the physical medium, Mind, and Matter:

> Thus...we arrive at the conclusion that Knowledge, Perception, Consciousness are messages or modes of communication between various selves - words as it were by which intelligences come to expression, and become known to each other and themselves. All Nature - all the actual world, as known to us or any being - we have to conceive as the countless interchange of communication between countless selves; or, if these selves are really identical, and the one Ego underlies *all* thoughts and knowledge, then the Subject and Object are the same, and the World, the whole Creation, is Self-revealment.[46]

Thus comes Carpenter to give voice to the questions we previously encountered in the Transhumanist vision of the Singularity:

> There is therefore, I say, a real universal Self, but there is also an elusive self. There are millions of selves which are or think thenselves separate. And over these we must delay. For to see the connection between them and the one Self is greatly important; and we may be sure that the illusive self is not for nothing; indeed the term 'illusive' may not after all be quite the right one to apply to it.
>   Let us ask two questions:-
>   1. How can the great Self also be millions of selves?,
>   2. If the great Self is within each of us, and the ego of every thought, why do we not know it so?[47]

Carpenter answers by reviving yet another image from the ancient metaphor, the idea of Man as Microcosm, for he evokes the image of the human body, with its countless cells all performing their own individual tasks at some rudimentary level of consciousness, and the greater human Self, neither arising from those cells, nor opposed to them, but somehow fully integral to them.[48] It is, once again, a both/and dialectic, not an either/or dialectic.

This leads Carpenter to an astonishing observation, one with deep applications to "the physics of the medium and consciousness" as we shall discover in the next chapter: "Some of us," he says, "who live in the midst of what we call Civilization simply live embedded among the thoughts of other people."[49] Civilization itself, in other words, has a kind of *psycho-* or *sociophysics*. Carpenter has here touched upon that deep mystery of the connection between individual consciousness, group consciousness, the physical medium, and the problem of theodicy, of good and evil. Volumes could, and have, been composed over this topic; we therefore leave it to the reader to divine the deep, and deeply dangerous, implications of what Carpenter has implied in his short and brilliant observation.

One clue, however, is afforded by Carpenter's speculations on the relationship of heredity to the Platonic idea of recollection (ἀναμνησις), for if there is a Self and many selves, there can, he speculates, arise selves somewhere between the two, "group" selves or "racial memories," which he explicitly speculates were the actual "scientific" bases for the ancient gods.[50] Had Carpenter known of the concept, he might also have had recourse to Carl Gustav Jung's idea of "archetypes". In any case, Carpenter, unlike other Uranian philosophers, also ties his understanding of the Platonic recollection of the "descent" of man as a hereditary phenomenon, to the idea that the primordial androgyny is also a *hereditary memory*.[51] In this, he may have been somewhat prophetic, as we shall see shortly. Additionally, in yet another prophetic observation, Carpenter observes that if consciousness of an individual human is an integrated phenomenon with that of the individual cells of his body, or in turn, that there is a kind of group hereditary consciousness, then at all levels - and not just at the level of the primary differentiation - consciousness is non-local.[52]

In touching on the idea of hereditary individual and group memory, Carpenter was extrapolating on ideas associated with the name of Darwin, only in this case, it is not Charles Darwin, but his grandfather, Erasmus Darwin, with whom we are concerned, for it is the elder Darwin who exercised such a hold on the imaginations of these alchemical-literary philosophers, from Percy Bysshe Shelley, to the Uranians, and even Oscar Wilde.

2. Erasmus Darwin, "The Temple of Nature," and the Evolutionary Algorithms of Differentiation

Erasmus Darwin, like Percy Shelley and Oscar Wilde, considered poetry and literature to be an alchemical technique whose ultimate utility was for the transformation of mankind. And like Shelley, he also knew that science was close to discovering the "algorithms of natural change," and embodied this worldview in a highly esoteric, and lengthy poem, *The Temple of Nature: or, the Origin of Society*, which, carried with it yet another subtitle: *A Poem, with Philosophical Notes*. It is heady, intoxicating and even mystical reading, for it combines the ancient esoteric metaphors, explaining them and extrapolating them on the basis of the then-modern science.

And like Shelley, Wilde, and Carpenter, his views were prophetic of later scholars of the esoteric. Consider, for example, these short lines on the talismanic and magic nature of Egyptian hieroglyphs:

Unnumber'd aisles connect unnumber'd halls,
And sacred symbols crowd the pictur'd walls,
With pencil rude forgotten days design,
And arts, or empires, live in every line.[53]

If one did not know better, one might think one was reading the thoughts of the twentieth century "alternative Egyptologist" Rene Schwaller DeLubicz, except expressed in the concision of art and poetry:

Hieroglyphic writing has the advantage over the Hebrew of utilizing images that, without arbitrary deviations, *indicate the* qualities and *functions in each sign.*

Cabalistic writing maintains secrecy but offers a clue by accentuating the principal idea, inexpressible by fixed concepts. It always employs a form of transcription *with several possible meanings*, using an ordinary fact as a **hook** to catch the thought....[54]

For our purposes, it is also important to note that Darwin believed the origin of the Greek mystery schools was in fact in Egypt,[55] a point of view reflecting his deep familiarity with esoteric traditions, and in fact, a point of view that has re-emerged in modern scholarship only recently.

Consequently it should come as no surprise that Darwin reproduces the Topological Metaphor, complete with references to its first differentiation or "the First Event,"[56] the birth of everything from stars:

Ere Time began, form flaming Chaos jurl'd,
Rose the bright spheres, which form the circling world;
Earths from each sun with quick explosions burst,
And second planets issued from the first.
Then, whilst the sea at their coeval birth,
Surge over surge, involv'd the shoreless earth;
Nurs'd by warm sun-beamns in primeval caves
Organic Life began beneath the waves.[57]

This is yet another view in line with modern cosmological physics ideas.

For Darwin, however, there is an implicit connection between these proto-evolutionary ideas then entering the scientific consciousness, and alchemy, for immediately after these lines he refers to the "First HEAT from chemic dissolution spring, and gives to matter its eccentric wings,"[58] an oblique reference to the transformative alchemical fires used to produce the Philosophers' Stone. In fact, reading Darwin closely, he is really maintaining that *all* matter, given its origins from "the Deity," is really a transmutating, information-creating Philosophers' Stone. All of this is an on-going, spontaneous process of differentiation and an alchemical "virgin birth":

> Hence without parent by spontaneous birth
> Rise the first specks of animated earth:
> From Nature's womb the plant or insect swims,
> And bungs or breathes, with mycroscopic (sic) limbs.[59]

But what has all this to do with Egypt and the Topological Metaphor?

Darwin's answer occurs in some breathtaking, and beautiful, lines that link Egypt, alchemy, and the idea of *differentiation*:

> So erst, ere rose the science to record
> In letter'd syllables the volant word;
> Whence chemic arts, disclosed in pictured lines,
> Liv'd to manking by heiroglyphic signs;
> And clustering stars, pourtray'd(sic) on mimic spheres,
> Assumed the forms of lions, bulls, and bears;
>   -So erst, as Egypt's rude designs explain,
> Rose young DIONE from the shoreless main;
> Type of organic Nature! source of bliss!
> Emerging Beuty from the vast abyss!
> Sublime on Chaos borne, the Goddess stood,
> And smiled enchantment on the troubled flood;
> The warring elements to peace restored,
> And young Reflection wondered and adored.[60]

Note that Darwin has faithfully produced another aspect of the Metaphor, namely, its both/and nature to live in a theistic interpretation, and in an atheistic one, by naming the primordial Nothing from which all else derives "Chaos" and "vast abyss." In this, Darwin is anticipating the developments of modern non-equilibrium thermodynamics and chaos theory, which show the nature of chaotic systems to "self-organize." Darwin, in other words, is maintaining that a sophisticated physics was once behind the "chemic" that is to say, "alchemic arts" of Egypt, and he is also perceiving quite accurately the "both/and" dialectic at work in the Metaphor.

Darwin also reproduces the doctrine that man is a microcosm, but in a subtle fashion, by pointing out that since man is *the end result of the process of differentiation, he is, in some sense, connected to all existence:*

> Imperious man, who rules the bestial crowd,
> Of language, reason, and reflection proud,
> With brow erect, who scorns this earthy sod,
> And styles himself the image of his God;
> Arose from rudiments of form and sense,
> An embryon point, or microscopic ens.[61]

All of this, Darwin states is a process of endless alchemical differentiation:

> Contractile earths in sentient forms arrange,
> And Life triumphant stays their chemic change.[62]

While it may be argued that "chemic" may mean "chemical" in the standard scientific sense (and indeed Darwin does use the term that way) the esoteric influence on his poem is no longer in doubt when Darwin turns to a consideration of man and to his primordial masculine alchemosexual androgyny:

> - HENCE ere Vitality, as time revolves,
> Leave the cold organ, and the mass dissolves;
> The Reproductions of the living Ens
> From sires to sons, unknown to sex, commence.
> New bugs and bulbs the living fibre shoots
> On lengthening branches, and protruding roots;
> Or on the father's side from bursting gland
> The adhering young its nascent form expands;
> In branching lines the parent-trunk adorns,
> And parts ere long like plumage, hairs, or horns.[63]

To put it briefly, Darwin is reproducing the old esoteric tradition that the male, the masculine, is the differentiating, creative force.

Only *later* in his poem does this "male androgyny" separate into the two sexes.[64] Darwin even attributes the rabbinical idea that mankind was created as a "male androgyne" to an Egyptian origin in one of his many footnotes. The Genesis account of the creation and of man, he notes,

...originated with the magi or philosophers of Egypt, with whom Moses was educated; and that this part of the history, where Eve is said to have been made from a rib of Adam, might have been an hieroglyphic design of the Egyptian philosophers, showing their opinion that mankind was originally of both sexes united, and was afterwards divided into males and females; an opinion in later times held by Plato, and I believe by Aristotle, and which must have arisen from profound inquiries into the original state of animal existence.[65]

But all this was the "alchemical science" of the early nineteenth century, the science of Shelley's *Frankenstein* and Erasmus Darwin's *Temple of Nature*. What, if anything, does contemporary genetics have to say about the matter?

C. THE BEHAVIORAL GENETICS CONFIRMATION, OR FALSIFICATION?

As it turns out, it has a great deal to say about the matter, and in saying it, it seems oddly enough both to confirm certain salient aspects of the ancient metaphor, and to falsify others. For example, the Oxford geneticist, Brian Sykes, comments at some length on what we have been calling the "embryonic androgyny:"

For the first sex weeks of development, human embryos destined to become male and female are indistinguishable from one another. We know, of course, that one has two X-chromosomes and the other an C- and a Y-chromosome, but up to this stage of development there is no way, short of a genetic test, of telling them apart. *They both have a pair of unisex gonads and two sets of primitive tubing called the Wolffian and Mullerian duct systems, named after their eponymous discoverers. During the seventh week of gestation the master gene, embedded in the Y-chromosome, is switched on in the male* -- but only for a few hours. The SRY protein, built to the precise orders of the sex gene, peels off the production line and heads off to activate other genes on several different chromosomes. From there, these genes trip a succession of genetic relays and under the influence of these secondarily activated genes, his unisex gonads begin to develop into testes which, before long, start to produce two different hormones. One is the descriptively named anti-Mullerian hormone, or AMH, which effectively destroyed the Mullerian duct system.

The other hormone produced by the embryonic testis is much better known. It is testosterone. AT this early stage in the growing male embryo, testosterone prevents the other system or primitive tubes, the Wolffian ducts, from being destroyed as they are in women. As time passes the Mullerian ducts disappear and the Wolffian ducts begin to expand to form the components of the internal male sexual organs - the prostate gland and seminal vesicles, and the *vas deferens* which connects them. Finally, some of the testosterone is converted into a high-octane form of the hormone - called dihydrotestosterone - and this organizes the growth of the external genitalia. Folds of tissue surround the urethra and form the penis, while nearby other tissues swell and fuse together to become the scrotum into which the testes eventually descend.

Female embryos, oblivious to the genetic stirrings on the Y-chromosome because they don't have one, proceed along their developmental pathway undisturbed by the irresistible hormonal signals coursing through their male counterparts.[66]

To put all of this as plainly as possible, modern genetics falsifies the ancient metaphor in one significant way, in that it is the *female* rather than the *male* which is the "default" setting for the program of human embryonic development.[67]

But is also *confirms* the ancient metaphor in another significant way, in that it is the *male* that is a *"special program" of differentiation*, for without the Y chromosome, development would proceed along the lines of the female. We cannot help but recall the fact that in ancient Mesopotamian lore, the "gods" engineered mankind by a chimerical mixture between a proto-human "female" donor and a "god" male donor.[68]

But the most astonishing genetic research that may confirm the intuitions of the nineteenth century Uranians, comes in the form of apparent genetic and neuro-physiological differences between most humans and homosexuals. Brian Sykes sums it up as follows:

The scientific literature on the biological basis for sexual orientation is a battleground of claim and counterclaim. With that proviso, here are some of the possibilities. For the most part they revolve around the notion that, just as male anatomy develops in the foetus under the direction of testosterone away from a feminine developmental pathway, so development of the male brain is a diversion from an otherwise female plan. Under this scheme male homosexuality is explained by a hitch in the transition to the male pattern. The anatomy of men's and women's brains is surprisingly similar, even though they act and think so differently, and only after a lot of detailed comparisons were any consistent differences found between the two. One of them lies within the hypothalamus, and its detailed description is 'the central subdivision of the bed nuclear of the stria terminalis' or BST for short. It would take another chapter to explain just what this is, but all we need to know here is that the BDT is two and a half times bigger in males than females, that it has plenty of sex hormone receptors and that it is wired into another brain structure, the small, almond-shaped amygdala. The amygdala is like a crossroads in the brain" the hub of an interconnecting network of neurological pathways and the seat of many of our emotions. The clue to the BST's association with gender identity and sexual orientation came when a team of Dutch scientists from Amsterdam conducted post-mortem examinations of the brains of six male-to-female transsexuals, men who had from childhood onwards had a strong feeling that they had been born the wrong sex. The Dutch team found that the BSTs of these men were much more similar in size and structure to those found in the typical female brain than to those of a man's brain.[69]

In other words, for whatever reason, it would appear that in some humans the sexual development of the gonads takes one pathway, the male one, while the neuro-physiological development takes the other, the female. Genetics may be confirming the old adage that homosexuals are indeed "wired differently," and that their orientation was never a choice.

All of this, of course, as Sykes himself states, has not been without controversy, and we point out that genetics, while providing ethical *clues*, has not yet provided ethical *answers*, for it could be hypothetically argued that the broad culture will compel geneticists to conclude that these developmental pathways are "abnormalities" and therefore to seek "genetic cures" of the "problem." And as we shall shortly see, religion - specifically in the form of Yahwism - once again enters the picture.

We return now to yet another reconsideration of the metaphor, something we first noted in our book *The Grid of the Gods*:[70] why is it, in so many ancient systems, that it is the male or masculine that is associated with *differentiation*, and the feminine or female, with *sameness or union?* The answer should now be evident, for it would appear that the metaphor was initially constructed in a scientific culture that knew that biologically it was the male that indeed was responsible for the primary differentiation that all humans are familiar with: sexual differentiation, for the "default" program is set to "female."

Thus the profundity and sophistication of the metaphor, and we speculate that the reason why so many cultures construed the primordial androgyny as a *masculine* adnrogyny - *and its ability to stand the test of time* - is revealed, for the more science advances, the more true the metaphor apparently seems. For example, the initial undifferentiated $\varnothing$ is a true, or "pure" androgyny, perhaps more feminine than masculine in its unending sameness, yet masculine, in that it can differentiate itself into $\varnothing_x$ and $\varnothing_y$. And yet, even in that act, it remains androgynous, and as equally feminine in that it reproduces only more versions of nothing, of itself. This both/and quality would, to an ancient scientifically advanced culture have been exemplified in males, since males carry both sexes. The "default" is set to "female," to reproduction of sameness, to $\varnothing$, it requires *special algorithmic programming to produce the male, to produce differentiation.* The resulting first differentiations, $\varnothing_x$ and $\varnothing_y$ are likewise androgynous, since they are masculine as differentiated (the "x" and the "y") and feminine since they are the same (the $\varnothing$).

We now have an approximate basis to understand why, even within the little-known Jewish and Christian esoteric traditions, some commentators believe the division of the sexes to have been accomplished in prevision of, or as a result of, the Fall, for in order to man to fall, the *total* man - the differentiated person (Adam, the masculine) and the undifferentiated nature (Eve, the feminine) - has to fall. Or to put it as bluntly as possible, the Fall was accomplished when the "x's" and "y's" of the symbolic notation ceased to indicate aspects of a real androgyny, and became the symbols of real sexual distinctions.

Thus, while we are far from claiming that modern science is confirming the ancient metaphor of androgyny, we do maintain that, for the *esotericists*, it would appear to do so, and be so interpreted by them. But before we can explore a few examples of how this esoteric tradition survived within the Judeo-Christian tradition, it is worth taking a brief excursion into shamanism.

D. Shamans, Drugs, DNA, and Androgyny

Jeremy Narby was an anthropologist who decided to study the use of hallucinogenic drugs in the shamanistic cultures of Native American tribes in South America. In doing so, what he discovered astonished him. The mystery expressed itself for Narby, as it so often does for people involved in such research, as a conflict between the apparently and self-evidently advanced knowledge of such cultures, and the primitive state of their existence. Narby puts his perplexity this way:

The main enigma I encountered during my research on Ashaninca ecology was that these extremely practical and frank people, living almost autonomously in the Amazonian forest, insisted that their extensive botanical knowledge came from plant-induced hallucinogens. How could this be true?

The enirma was all the more intriguing because the botanical knowledge of indigenous Amazonians has long astonished scientists. The chemical composition of ayahuasca is a case in point. Amazonian shamas have been preparing ayahuasca for millennia. The brew is a necessary combination of two plants, which must be boiled together for hours. The first contains a hallucinogenic substance, dimethyltriptamine, which also seems to be secreted by the human brain; but this hallucinogen has no effect when swallowed, because a stomach enzyme called monamine oidase blocks it. The second plant, however, contains several substanhces that inactivate this precise stomach enzyme, allowing the hallucinogen to reach the brain. Thyc sophstication of this recipe has prompted Richard Evens Schultes, the most renowned ethnobotanist of the twentieth century, to comment: "One wonders how peoples in primitive societies, with no knowledge of chemistray or physiology, ever hit upon a solution to the activation of an alkaloid by a monoamine oxidase inhibitor. Pure experimentation? Perhaps not. The examples are too numerous and may become even more numerous with future research."

So here are people without electron microscopes who choose, among some 80,000 Amazonian plant specials, the leaves of a bush containing a hallucinogenic brain hormone, which they combine with a vine containing substances that inactivate an enzyme of the digestive tract, which would otherwise block the hallucinogenic effect. And they do this to modify their consciousness.

It is as if they knew about the molecular properties of plants *and* the art of combining them, and when one asks them how they know these things, they say their knowledge comes directly from hallucinogenic plants.[71]

Narby decided to test the drug himself, for the shamans reported seeing serpents, spiraling ladders, and a host of other bizarre visions.

His own use of the drug, under the direction of an Ashaninca shaman, led him to formulate a thesis that would have delighted Edward Carpenter, and some modern transhumanists:

My investigation had led me to formulate the following working hypothesis: In their visions, *shamans take their consciousness down to the molecular level and gain access to information related to DNA, which they call "animate essences" or "spirits."* This is where they see double helixes, twisted ladders, and chromosome shapes. This is how shamanic cultures have known for millennia that the vital principle is the same for all living beings and is shaped like two entwined serpents (or a vine, a rope, a ladder...).... The myths of these cultures are filled with biological imagery.[72]

In other words, such ancient images as the caduceus, the entwined serpents, ladders to heaven, are the encoded science of the DNA double helix, which all life shares.[73]

But more importantly, Narby is implying that through the mechanism of drug use, shamans integrate their consciousness with the rudimentary consciousness of the cells of their body, in yet another manifestation of the both/and dialectic we encountered previously with Carpenter. In one such image of entwined serpents, from the Desana Shamanism, the two serpents represent "a female and a male principle, a mother and a father image, water and land.... in brief, they represent a concept of binary opposition which has to be overcome in order to achieve individual awareness and integration."[74] The reintegration, of course, is an androgyny.

The connection to a scientific sophistication exhibited by these ancient cultures only became more bizarre for Narby, for he noted that Autralian aborigines also told the same story as the Amazonian Indians half a world away, and here, the image is even more striking:

This time is was Australian Aborigines who considered that the creation of life was the work of a "cosmic personage related to universal fecundity, the Rainbow Snake," whose powers were symbolized *by quartz crystals.* It so happens that the Desana of the Columbian Amazon also associate the cosmic anaconda, creator of life, with a quartz crystal....

How could it be that Australian Aborigines, separated from the rest of humanity for 40,000 years, tell the same story about the creation of life by a cosmic serpent associated with a quartz crystal as is told by ayahuasca-drinking Amazonians?[75]

And why associate all of this with *a crystal,* as if implying, once again, the alchemical idea of "mineral man"?

Narby dug into the phenomenon of "biophotons," the small photons of light emitted by living cells, and made yet another astonishing discovery:

> One thing had struck me as I went over the biophoton literature. Almost all of the experiments conducted to measure biophotons involved the use of *quartz*. As early as 1923, Alexander Gurvich noticed that cells separated by a quartz screen mutually influences each other's multiplication processes, which was not the case with a metal screen. He deduced that cells emit electromagnetic waves with which they communicate. It took more than half a century to develop a "photomultiplier" capable of measuring this ultra-weak radiation: the container of this device is also made of quartz.[76]

After pointing out that various shamanistic cultures play a significant role in their rituals,[77] Narby points out that DNA is itself an aperiodic crystal:

> The four DNA bases are hexagonal (like quartz crystals), but they each have a slightly different shape. As they stack up on top of each other, forming the rungs of the twisted ladder, they line up in the order dictated by the genetic text. Therefore, the DNA double helix has a slightly irregular, or aperiodic, structure. However, this is not the case for the repeat sequences that make up a full third of the genome.... In these sequences, DNA becomes a regular arrangement of atoms, a *periodic* crystal - which could, by analogy with quartz, pick up as many photons as it emits. The variation in the length of the repeat sequences...would help pick up different frequencies and could thereby constitute a possible and new function for a part of "junk" DNA.[78]

In other words, there was a direct analogy between crystals and life, specifically quartz, which gives yet another window of confirmation into the ancient metaphor of "alchemo-mineral" man, and also may have something to do with why so many cultures the world over built massive pyramids out of materials embedding innumerable quartz crystals in its stones.

This strange and ancient imagery led Narby to make a breathtaking hypothesis, which we cite here in order that its full implications may sink in:

> ... I did know that DNA was an aperiodic crystal that traps and transports electrons with efficiency and that emits photons(in other words, electromagnetic waves) at ultraweak levels currently at the

limits of measurement - and all this more than any other living matter. This led me to a potential candidate for the transmissions: the global network of DNA-based life.[79]

In other words, viewed a certain way, the fact of biophotonic energy means that the sum total of human, plant, and animal DNA on the earth acts as a huge transmitter/receiver of signals, the ultimate "interface" between consciousness and the physical medium, and giving, perhaps, a scientific basis to the speculations of Carpenter that there can arise a kind of "group consciousness" multiplier effect.

At this juncture, it is worth noting that Carpenter points out that among similar "primitive" Indian cultures in *North* America there were words for homosexual individuals in those societies that were called "half-men half women,"[80] and that, in keeping with these cultures' views on the primordial androgyny, such people were not ostracized, but rather, revered.

For these primitive cultures, the serpent image thus constitutes an image both of life, and of androgyny,[81] and not surprisingly, it is the arrival of Yahwism that *inverts* the image and overturns the old symbolism:

> (Joseph) Campbell dwells on two crucial turning points for the cosmic serpent in world mythology. The first occurs "in the context of the patriarchy of the Iron Age Hebrews of the first millennium B.C., (where) the mythology adopted from the earlier neolithic and Bronze Age civilizations...became inverted, to render an argument just the opposite of its origin." In the Judeo-Christian creation story told in the first book of the Bible, one finds elements which are common to so many of the world's creation myths" the serpent, the tree, and the twinbeings; but for the first time, the serpent, "who had been revered in the Levant for at least seven thousand years before the composition of the Book of Genesis," plays the part of the villain. Yahweh, who replaces it in the role of the creator, ends up defeating "the serpent of the cosmic sea, Leviathan."
>
> ...
>
> At this point, I wrote in my notes, "These patriarchal and exclusively masculine gods are incomplete as far as nature is concerned. DNA, like the cosmic serpent, is neither masculine nor feminine, even though its creatures are either one or the other, or both. Gaia, the Greek earth goddess, is as incomplete as Zeus. Like him, she is the result of the rational gaze, which separates before thinking, and is incapable of grasping the androgynous and double nature of the vital principle."[82]

It should come as no surprise then, that some of the loudest objections against recent genetics and neuro-physiological studies are coming from the Yahwist tradition, and to these final considerations we now turn.

E. Objections: Religion and Yahwism Again

In a lengthy article posted on the internet, Dr. Peter R. Jones summed up the evangelical Christian response to all these types of developments in the expected fashion:

> ...the urgency of the situation for Bible-believing scholars is not merely the pressing need for a scholarly ethical response to an unfortunate moral aberration. The contemporary appearance of a homosexual movement says something about the particular times in which we live, granted both that pagan spirituality is enjoying a popular revival, and that throughout the Bible, Sodom and Gomorrah have always served as the symbol for endtime pagan idolatry, ultimate moral disintegration and eschatological divine judgement. *The subject, in its spiritual, religious and and even eschatological dimensions, needs to be treated and debated among us, not simply as an unfortunate social deviation or ephemeral social fad, but as a cutting-edge component of a rising, all-encompassing, religious world view that is diametrically opposed to the world view of Christian theism.*[83]

In his article, Jones traces the influence of these doctrines to ancient Christian Gnosticism, and to mediaeval alchemy:

> Later in the second and third centuries of the Christian church, the Gnostics were credited by their adversaries with mystery celebrations involving carnal knowledge. The charge is credible because "Christian" Gnosticism was the attempt to Christianize pagan spirituality, even to the point of adopting some form of androgyny. Hippolytus (AD 170-236) reports that one particular Gnostic sect, the Naasenes, who worshipped the Serpent (Naas in Hebrew) of Genesis, attended the secret ceremonies of the mysteries of the Great Mother in order "to understand the 'universal mystery.'" Like modern syncretists who are encouraged to cross over into other religions, the Gnostics believed religious truth was one, to be found everywhere, and so they crossed over into pagan spirituality as a matter of religious principle. The most explicit testimony is from Irenaeus who says: "They prepare a

bridal chamber and celebrate mysteries." A homosexual encounter is perhaps insinuated in the "Secret Gospel of Mark." At the very least, the final logion 114 of the Gospel of Thomas appears to be an invitation to spiritual androgyny. All this would justify the judgment of Burkhart that "certain Gnostic sects seem to have practiced mystery initiations, imitating or rather outdoing the pagans…"

There is good reason to believe that a form of ancient Gnosticism, namely Hermeticism, survived and influenced the Medieval West through the mystical spirituality of Alchemy. This variant Egyptian version of Gnosis saw in Hermes the divine interpreter whose secrets enable Man to pass through various levels of reality, thus making esoteric transmutations possible. The spiritual alchemist became an initiate, one "who knows," as the ancient Gnostics "knew." Like Hermes, the alchemical Mercurius was understood as a kind of divine "other" who would intervene by affecting the resolution of opposites. While no explicit sexual perversion is promoted, joining of the opposites or union was frequently imaged as a hieros gamos, a holy marriage, the fruit of which is called "the Philosopher's Stone." This "fruit" is sometimes called "the child of the work" which is presented as the Hermetic Androgyne, under the rubric "Two-in-One." At the very least we have to reckon here with a spiritualized form of what 'Eliade calls "ritual androgynisation."[84]

The bow to early Christian traditions, however, is, we are bold to suggest, very incomplete, for the primordial androgyny was very much a part of more mainstream Jewish rabbinical tradition, and Christian patristic tradition.

For example, the famous Christian saint, Maximus the Confessor (ca 580-662), wrote that in the union of Christians in Christ, that the difference between male and female was "mystically abolished,"[85] a gloss on the fact that in Galatians 3:28, the unification of opposite sexes is used as "a prime symbol of salvation:"[86] in Christ "there is no Jew nor Greek, there is no slave nor free, there is no male and female." The androgyny, in other words, becomes more subtle:

However many varied resonances the early Christian ritual clothing language may evoke, it is most fundamentally related to a particular myth. The "new man" symbolized by the clothing is the man who is "renewed according to the image of his creator" (Colossians 3:10; cf. Ephesians 4:24). The allusion to Genesis 1:26-28 is unmistakable; similarly, as we noted earlier, Galations 3:28 contains a reference to the "male and female" of Genhesis 1:27 and suggests that somehow the act of Christian initiation reverses the fateful division of Genesis 2:21-22. Where the image of God is restored, there, it seems, man is no longer divided - not even by the most fundamental division of all, male and female.[87]

Notice the reference to Genesis 1:26-27, and Genesis 2:21-22. What is going on here is that within rabbinical tradition, there is a gloss on verse 1:27 of Genesis from the Babylonian Talmud which interprets the verse to mean "A male with corresponding female parts created He *him*,"[88] a gloss echoed in the Palestinian Talmud as "male with female parts he created *them*." These reading were reflected in the Spetuagint Greek translation of the Old Testament, where the word "them" occurs in the *masculine* gender.

Reflecting on this tradition, the ninth century Christian philosopher, John Scotus Eriugena(815-877), greatly influenced by Maximus the Confessor, made the following comment:

Nor is Scripture silent about this: For concerning the fact that, immediately after the transgression, human nature, *which before its sin had been simple, was after its fall divided into two sexes* it says: "And they sewed fig-leaves together..."[89]

In other words, there appears to be a kind of schizophrenia within Yahwist religion, between a primordial *and eschatological* androgyny, and the contemporary "dispensation" in which its manifestations are to be fought, denied, or suppressed.

The schizophrenia is perhaps best illustrated by a curious theophany of Yahweh to Moses recorded in the book of Exodus, a theophany whose baldly "alchemosexual" implications can hardly be overlooked:

And he(Yahweh) said, "Thou canst not see my face: for there shall no man see men, and live. And the LORD said, Behold, there is a place by me, and thou shalt stand upon a rock: and it shall come to pass, while my glory passeth by, that I will put thee in a cleft of the rock, and will cover thee with my hand while I pass by: and I will take away mine hand, and thou shalt see my back parts; but my face shall not be seen.[90]

The sexual imagery of Moses standing in the cleft of a rock while Yahweh exposes his back parts is perhaps one of the most disconcerting in all the strange imagery we have encountered in this book, for if one takes Yahweh as a purely masculine entity, the moral implications implied are disconcerting indeed.

But the question is, *should* Yahweh be construed in that fashion? Once again, Uranian Edward Carpenter pointed out one of those little-known things, little known, because it is an obvious thing hidden in plain sight, in this case, in the characters of the Hebrew text, regarded as sacrosanct by Judaism and certain strains of Protestant Christianity, for "the two words of which Jehovah is composed make up the original idea of male-female of the birth-originator. For the Hebrew letter Jod (or J) was the *membrum virile*, and Hovah was Eve, the mother of all living, or the procreatrix Earth and Nature."[91] In other words, the very name Yahweh itself, read in the symbolism of the very Hebrew characters of which it was composed, was symbolically androgynous. The schizophrenia, we suggest, lies in the nature of the religion itself, with its claims to be an "absolute morality" on the one hand, and yet, with its own powerful androgynous images locked within one of its sacred texts, an image used, in fact, to portray the final eschatological transformation of man in a state where there is "neither male nor female."

***

What is one to make of all this?

We believe that modern science - at least in its biological component - is coming to a position where the strangest alchemical image of them all, that of androgyny itself, is found to have at least *some* basis in reality, and that this fact *perhaps* indicates that it was born in High Antiquity, and in a similarly sophisticated scientific culture.

But is there a deeper aspect to this imagery, to the image of man as microcosm and of the universe as macanthropos? Is there, in fact, a physics component that lurks beneath the ancient metaphors, in addition to the biological one?

Brace yourself...

Endnotes

1   Arturo Schwarz, "Alchemy, Androgyny and Visual Artists," *Leonardo*, Vol, 13, pp. 57-62. www.jstor.org/pss/1577928.

2   Michelle Alford, "AFIRM Reduces Scar Formation with Their Advances in Regenerative Medicine," Scars1, http://www.scars1.com/news/ mainstory.cfm/55.

3   Martin Hutchinson, "Mixed-sex Human Embryo Created," July 3, 2002, http:// news.bbc.co.uk/2/hi/health/3036458.stm.

4   Maggie Fox, "US company claims cloned humans, made stem cells," Jay 17, 2008, http://www.reuters.com/article/2008/01/21/ idUSN1721774620080121.

5   http://www.ornl.gov/sci/techresources/Human_Genome/elsi/ behavior.shtml.

6   Ibid.

7   Alice Domurat Dreger, *Hermaphrodites and the Medical Invention of Sex* (Harvard University Press, 2000), p. 25.

8   Ibid., p. 154.

9   Ibid., p. 26.

10   Ibid., p. 25.

11   Ibid.

12   In some cultures, the Lakota Sioux of the Dakotas, for example, the occurrence of male homosexuality or hermaphroditism was actually viewed as a *gift*, since it was believed that the consciousness of such individuals was simultaneously closer to God, being both intuitive and nurturing (female), and analytical and rational (male).

13   Dreger, op. cit., p. 16.

14   Ibid., p. 24.

15   Ibid.,pp. 20, 29.

16   Ibid., pp. 1-2, 20. Dreger actually begins her book with a consideration of the French "Sophie," a women who upon gynecological examination proved to be a man, but who, due to her unusually ambiguous genitalia, had been raised by her parents as a girl. On pp. 19-20 Dreger cites the famous case of Alexina/Abel Barbin, a British hermaphrodite, whose ambiguous genitalia - and Dreger actually reproduces the medical drawings of the same on p. 19, and we can attest, they are truly hermaphroditic - caused similar problems.

17   Ibid., p. 85.

18   Ibid.

19   Ibid.

20   Ibid., p. 83.

21   Ibid., p. 79.

22   Ibid., p. 35.

23   Ibid., p. 34. See also p. 155.

24   Ibid.

25   Ibid., pp. 69-70.

26   Ibid., p. 136.

27   It should be pointed out that Mayne and other Uranians were quick to note that hermaphroditism of the sort being studied by 19th century medicine was *not*, in their view, equivalent to male or female homosexuality, since most homosexuals were "gonadically" either male or female. (cf. p. 134). What they were arguing, however, was that homosexuality, and even bi-sexuality, were themselves possibly equally hereditary, and neither more nor less "monstrosities" or "pathologies" than was hermaphroditism, but rather, possibly manifestations in adult form of the primordial androgyny, itself manifested in the "embryonic androgyny."

28   Ibid., p. 138.

29   Ibid., p., 37.

30   Edward Carpenter, *The Art of Creation: Essays on the Self and Its Powers* (Bibliolife

reprint of the George Allen-Ruskin House edition of 1904), p. 58.

31 Edward Carpenter, *The Intermediate Sex: A Study of Some Transitional Types of Men and Women* (Kessinger Publications Reprint of the George Allen & Co. Ltd. edition of 1912), pp. 34-35.

32 Ibid., p. 56.

33 Ibid., p. 72.

34 Edward Carpenter, *The Art of Creation*, pp. 4-5, emphasis added.

35 Ibid., pp. 115-116, boldface emphasis added.

36 Ibid.,p. 35.

37 Ibid., p. 34.

38 Ibid.,p. 42.

39 Ibid., p. 33.

40 Ibid., p. 46.

41 Ibid., pp. 47-48.

42 Ibid., pp. 48-49.

43 Ibid., p. 51.

44 Ibid., pp. 54, 30-31.

45 Ibid., p. 58.

46 Ibid., p. 44.

47 Ibid., pp. 71-72.

48 Ibid., p. 85.

49 Ibid., p. 25.

50 Ibid., p. 123.

51 Ibid., p. 131.

52 Ibid., p. 139. Carpenter's actual statement of this point is extraordinarily subtle: "If the existence of race-memories, and of feelings and visions accompanying them, is allowed at all, it would seem that these things must belong in some degree to the consciousness of the race, to a less individual *and local consciousness than the ordinary one.* The terms 'mania,' then, or 'ecstasy,' which would indicate the passing out from the ordinary consciousness (into the racial or celestial, according as we adopt the modern or Platonic view), would seem quite appropriate."

53 Erasmus Darwin, M.D.F.R.S., *The Temple of Nature; or, the Origin of Society: A Poem, with Philosophical Notes* (British Library Reprint of the New York, T. and J. Swords Edition, 1804), pp. 12-13.

54 R.A. Schwaller de Lubicz, *The Egyptian Miracle: An Introduction to the Wisdom of the Temple* (Inner TrRaditions, 1985), p. 8, italicized emphasis added, boldface emphasis in the original. Cited in Joseph P. Farrell and Scott D. de Hart, *The Grid of the Gods* (Adventures Unlimited Press, 2011), p. 276.

55 Erasmus Darwin, *The Temple of Nature*, p. 16.

56 Ibid., pp. 20-21.

57 Ibid., p. 21.

58 Ibid.

59 Ibid., p. 23.

60 Ibid., pp. 31-32.

61 Ibid., p. 27.

62 Ibid., p. 35.

63 Ibid., p. 43.

64 Ibid., pp. 46-47.

65 Ibid., p. 176.

66 Brian Sykes, *Adam's Curse: The Science that Reveals Our Genetic Destiny* (W.W. Norton & Company, 2004), pp. 72-73, emphasis added.

67 Ibid., p. 67.

68 Cf. Joseph P. Farrell, *Genes, Giants, Monsters, and Men* (Feral House, 2011), pp. 138-160.

69   Sykes, op. cit., p. 269. See also Rob Stein, "Brain Study Shows Differences Between Gays, Straights," *The Washington Post*, June 23, 2008, p. A12; and Dan Eden, "Homosexuality: Does it Have a Natural Cause?" at www.viewzone2.com'homosexualx.html, for a list of other inherited traits typical of homosexual men and women.

70   Joseph P. Farrell with Scott D. de Hart, *The Grid of the Gods* (Adventures Unlimited Press, 2011), pp. .

71   Jeremy Narby, *The Cosmic Serpent: DNA and the Origins of Knowledge* (Jeremy P. Tarcher/Putnam, 1998), pp. 10-11.

72   Ibid., p. 117, emphasis added.

73   Ibid., pp. 56-57, see also p. 29.

74   Ibid., p. 57.

75   Ibid., p. 64.

76   Ibid., pp. 128-129.

77   Ibid., p. 129.

78   Ibid., p. 130.

79   Ibid., pp. 109-110.

80   Edward Carpenter, *Intermediate Types Among Primitive Folk*, p. 68.

81   Narby, op. cit., pp. 65-66.

82   Ibid., pp. 66-67.

83   Peter R. Jones, "Americans for Truth about Homosexuality: Androgyny: The Pagan Sexual Ideal (It's Not About Civil Rights), http://americansfortruth.com/2007/03/09/androgyny-the-pagan-sexual-ideal-its-not-about-civil-rights/

84   Ibid.,

85   Maximus the Confessor, *Quaestiones ad Thalassium* 48 (Migne *Patrologia Graeco-Latina*) 90: 436A, cited in Wayne A. Meeks, "The Image of the Androgyne: Some Uses of a Symbol in Earliest Christianity," *History of Religions*, Vol. 13, No. 3 (University of Chicago Press, Feb 1974) 165-208, p. 165.

86   Meeks, op. cit., p. 166.

87   Ibid., p. 185.

88   Ibid., p. 185, no. 88.

89   John Scotus Eriugena, *Periphyseon(The Division of Nature)*, trans. I.P. Sheldon-Williams, revised by Jphn J. O'Meara (Dumbarton Oaks, 1987), pp. 188-189, emphasis added. Not surprisingly, Eriugena - again under influence from Maximus - reproduces the ancient metaphor of mankind as a *mircocosm* of the intelligible and sensible universe: "Man...was made in such an honorable position in created nature that there is no creature, whether visible or invisible, which cannot be found in him. By a marvelous union, he was compounded from the two universal parts of created nature, the sensible and intelligible, i.e., he was joined from the extremes of all creation." (John the Scot, *Periphyseon: On the Division of Nature*, ed. and trans Myra l. Uhlfelder [Bobbs-Merrill Company, Inc., 1976], p. 116.)

90   Exodus 33:20-23, Authorized Version.

91   Carpenter, *The Intermediate Sex*, p. 74.

⚜ Twelve ⚜

EPILOGUE IS PROLOGUE:
MICROCOSM AND MEDIUM,
THE ANTHROPIC COSMOLOGICAL PRINCIPLE IN PHYSICS

∴

"When (Percy Bysshe) Shelley reached Oxford, he poured out his thoughts concerning the possible uses of heat and combustion to transform matter, produce food, and eliminate starvation and slavery. Walker helped Shelley obtain a solar microscope, which used sunlight to project an enlarged microscopic image in a darkened room. It became a favorite possession at Oxford and later. As late as 1814, it was observed that 'Shelley makes chemical experiments.' The ideas of energy introduced by Walker became features of Shelley's poetics of "visionary physics," including chaos theory in contemporary physics."
—James Bieri[1]

"Get the maximum out of theology. Read it as you like."
—Anonymous

YOU PROBABLY HAVE NEVER HEARD OF Frank J. Tipler, and indeed, if you are not a physicist, there is no reason you would have. But he is the genuine article, a tenured professor of Mathematical Physics at Tulane University, and the author of three mind-expanding - or depending upon one's point of view, mind-boggling - treatises on the implications of recent developments in theoretical physics: *The Anthropic Cosmological Principle* with co-author John D. Barrow, and two sequels of his own, *The Physics of Immortality*, and *The Physics of Christianity.* Our concentration here in this short epilogue will be on the first book with co-author John D. Barrow, *The Anthropic Cosmological Principle.*

We thus end, where we began, with physics, and the "topological metaphor" of the physical medium. Obviously, given Tipler's vast output - the combined total of his three books is well over a thousand pages - we can but briefly survey his work here, highlighting those portions of it that appear to vindicate the ancient "topological metaphor" to a certain extent, leaving a detailed exposition to future works.

It will be recalled that there were three components of the metaphor that we have encountered along the way: (1) the idea of the universe as a "great man" or "makanthropos"; (2) the idea of man as a "little universe" or "microcosm," standing exactly mid-way between all possible binary polarities, and (3) the idea of the original "Nothing" as a kind of "androgyny," a fusion of various distinct polarities. The first two components are embodied in modern theoretical physics' conceptions as the "Anthropic Cosmological Principle," the modern version of Vitruvian Man and the old adage "Man is the measure of all things, of things that are, that they are, and of things that are not, that they are not;" or, to put it in terms of the Masonic symbolism of compass and square - and the physics implications that those symbols imply - man is, in this view, the Grand Architect of the Universe.

The principle, in all its four versions, is based on a certain underlying principle of quantum mechanics, and the role of the physical observer, in it: the Uncertainty Principle. Briefly put, this principle states that one cannot measure the position of an electron at the same time one measures its velocity. In other words, the experimenter himself pre-determines to a great extent the outcome of an experiment by the *selection* of what it is he wants to observe, and to that extent, determines reality itself. It does not take much further thought to see that pressed to its limits, this means that there is a kind of direct "interface" between consciousness and the physical medium, manifest in the principle of *selection* of what it is that is to be measured. This has led some physicists into the realm of philosophical speculation, a speculation no longer strictly metaphysical in the traditional sense, for now it is backed up by equations and experiment, and the result of these inquiries has been the formulation of the Anthropic Cosmological Principle among some physicists, which, breifly stated, is that "the basic features of the Universe, including such properties as its shape, size, age and laws of change, must be *observed* to be of a type that allows the evolution of observers."[2] The universe, in other words, is "hard-wired" to produce observers, i.e., intelligent life. As noted previously, this Anthropic Principle comes in four versions, each of which stresses, to varying degrees, the centrality of life, and of physical observers - man - of the universe:

1) *The Weak Anthropic Principle:* Barrow and Tipler state this, the weakest version, of the Anthropic Cosmological Principle as follows: "The observed values of all physical and cosmological quantities are not equally probable but they take on values restricted nby the requirement that there exist sites where carbon-based life can evolve and by the requirement that the Universe be old enough for it to have already done so."[3] This is a precise, though somewhat confusing way, of saying simply that "those properties of the Universe we are able to discern are self-selected by the fact that they must be consistent with our own evolution and present existence."[4]

2) *The Strong Anthropic Principle:* This version of the principle is basically the same, except that it is stated much more directly: "The Universe must have those properties which allow life to develop within it at some stage in its history."[5]

3) *The Participatory Anthropic Principle:* It is when one adds the principles of quantum mechanics, and in particular, the crucial centrality of the role of the observer, to the picture that the Strong Anthropic Principle changes character. Here it is worth citing Barrow and Tipler at length:

> ...the Inclusion of quantum physics into the SAP (Strong Anthropic Principle)Produces quite different interpretations. Wheeler has coined the title *Participatory Anthropic Principle* (PAP) for a second possible interpretation of the SAP: *Observers are necessary to bring the Universe into being.*[6]

Barrow and Tipler point out that if, in this version of the Anthropic Principle, at some point the Universe dies out before the sum total of observations made about it are able to have "any measurable... influence on the Universe in the large, then it is hard to see why it *must* have come into existence in the first place."[7] This leads to the following, and strongest, statement of the Anthropic Principle:

4) *The Final Anthropic Principle:* "Intelligent information-processing must come into existence in the Universe, and, once it comes into existence, it will never die out."[8]

It is with this last version that the transhumanist vision, the ancient metaphor, and modern physics, all come together, for as Barrow and Tipler observe,

Although the FAP (Final Anthropic Principle) is a statement of physics and hence *ipso facto* has no ethical or moral content, it nevertheless is closely connected with moral values for the validity of the FAP is the physical precondition for moral values to arise and to continue to exist in the Universe: no moral values of any sort can exist in a lifeless cosmology. Furthermore the FAP seems to imply a melioristic cosmos.[9]

With the idea that the Final Anthropic Principle implies a melioristic universe, i.e., a universe that mankind takes an active and participatory hand in improving, we have arrived at the basis in theoretical physics for the transhumanist agenda.

This places us chin to chin with the hidden, inherent "apocalyptic premise" within the Final Anthropic Principle, for the existence of intelligent observers posits a kind of "group consciousness" responsible for the coordination of all independent observations:

This line of speculation leads naturally to the fourth possibility, that there is some Ultimate Observer who is in the end responsible for coordinating the separate observations of the lesser observers and is thus responsible for bringing the entire Universe into existence.... The sequence of observers ... could continue to run until an observer O_i is reached in the future who, by his observation, coordinates two such sequences of observers. But O_i himself is part of another sequence which is joined further in the future by observer O_{i+1}. This joining of sequences of observers continues - and even includes the observations made by different intelligent species elsewhere in the Universe - until *all* sequences of observations by all observers of all intelligent species that have ever existed and will ever exist, or *all* events that have ever occurred and will ever occur are finally joined together by the Final Observation by the Ultimate Observer. He must be located at the final singularity in a closed universe, or at a future time-like infinity in an open universe.[10]

Note, that the temporal position of this "Ultimate Observer" is in *the future*, and that its function is as a coordinator of all observations of all individual consciousness; we are, in other words, once again back at the both/and nature of the metaphor. It is, so to speak, the result or sum total of all observations. The implications of this view for the age-old problem of theodicy are, of course, profound. It is the physics version of the transhumanists' goal of extending human consciousness to universal proportions and extent.

Modern physics even produces its own version of the Metaphor of creation out of Nothing, for it is possible to view "the whole universe to be a giant, quantum mechanical virtual fluctuation of the vacuum," if one but recall that the vacuum of quantum mechanics is "envisaged to be a sea of continually creating and annihilating particle-antiparticle pairs" that exist for a particular span of time.[11]

Even the ancient idea of man as a microcosm, positioned midway between the region of "spirit" and the region of "matter", as a kind of "common surface" of both regions,[12] is reproduced in analogous fashion, for one of the things that gave rise to the formulation of the Anthropic Principle in the first place was the realization by some physicists that "the mass of a human is the geometric mean of planetary and an atomic mass..."[13]

***

While this survey of the Anthropic Comsological Principle has been all too brief, one thing it does indicate is that, yet again, the ancient Topological Metaphor, with all its associated imagery of androgyny, of a descent of man from mineral to vegetable to animal, of man as microcosm and the universe as a "large man" or "makanthropos," is not without its possible foundations, either in biology, or now, as has been seen, in physics. We are therefore bold to suggest that this metaphor of ancient myths is born from a matrix of great and complex sophistication, because it *is* so very capable of being "reverse engineered" or rationalized in terms of the advances of modern science and technology. We are far from maintaining that our analysis is either complete, or thorough. This has been only an essay, an overview, each component of which could have occupied whole books in themselves. Nor are we maintaining that ours is the only or even best way so to "reverse engineer" and rationalize the metaphor. But it is, we believe, at least, one *possible* way to do so.

But those ancient stories and myths and metaphors contain also a profound warning, that the technologies and science underlying them are, indeed, in some sense divine, and threatening, that to wield them with anything less than love, and a profound and tender respect for all life, will lead to inevitable ruin. We are indeed entering a Hermetic, alchemical age, or perhaps *re*-entering it, a process begun long ago when the Renaissance Magicians transformed the culture with a supremely magical transformation. We have touched, all too briefly, on the implications of a physics of consciousness and its "group multiplier" effect, with all the deep and profound implications for traditional theodicy that this implies. As was seen, it is not only the physicists who see this; Percy Shelley prophesied of it, the Neoplatonists, Hermeticists, the ancient Hindu authors of the *Rig Veda*, and the modern Uranian Edward Carpenter, and so many more, all speculated on it, the Masons and Rosicrucians and other secret fraternities have entire traditions founded upon it.

Inevitably. the views of alchemists, geneticists, transhumanists, physicists, poets, and writers that we have reviewed here do not occur in a social or cultural vacuum; they are a threat to some, a gift to others; they will in one form or fashion, influence and affect us all. They gladden some, and sadden others. But we will all have to deal with them.

The "transhumanist grimoire" that we have surveyed here is not, obviously, without its implications for western culture, nor the underlying religious matrix in which these views came to be born and nurtured, and eventually, opposed. After all, it was the work of alchemists, hiding their secrets in dark primitive laboratories from the prying eyes and spies of Caliphs, Sultans, Popes, Bishops, Kings and Emperors, that continued and carried forward the old metaphor.

But the old monotheisms, the Three Great Yahwisms, are not simply going to roll over and die, for their very world view, their very culture, is at stake. They, after all, were themselves already formed as a vast inversion to the more ancient metaphor, preserving some aspects of it, rejecting others. They will not go down without a fight.

They call that fight Armageddon, the final clash of civilizations and ages. Their prophets have long foreseen it. And the elites who possibly spawned them, know this, and in their insane rush to the global transformation of man, have planned on it...

...but that's the subject of another book; Epilogue is always Prologue...

ENDNOTES

1   James Bieri, *Percy Bysshe Shelley: A Biography* (John Hopkins University Press, 2008), p. 51.

2   John D. Barrow and Frank J. Tipler, *The Anthropic Cosmological Principle* (Oxford University Press, 1986), p. 1.

3   Ibid., p. 16.

4   Ibid.

5   Ibid., p. 21.

6   Ibid., p. 22, emphasis in the original.

7   Ibid., p. 23.

8   Ibid.

9   Ibid.

10 Ibid., pp. 470-471.

11 Ibid., p. 440.

12 Manly P. Hall, *The Secret Teachings of All Ages*, Reader's Edition, p. 225: "Both God and man have a twofold constitution, of which the superior part is invisible and the inferior visible. In both there is also an intermediary sphere, marking the point where these visible and invisible natures meet."

13 Barrow and Tipler, op. cit., p. 12.

APPENDIX:

THREE ALCHEMICAL POEMS

ALCHEMY

Ancient wisdom, the poet's pen; a
Logical mystery in plain sight.
Chemically appearing in chimerical visions
    within the
Hermeticist's dream.
Esoteric advent, transubstantiated
Magic; driven to Babel, in
Yahweh's inferno.

IN PRINCIPIO

In Nothing. the limitless sea
Was sewn an androgynous Seed
Was spoken a soundless boundless Word
Ο Λογος ην εν αρχη
to summon forth endless varieties
of Itself in limitless nothings
and differentiation of multibody non-linearities
A primal function, Verbum, Word, Λογος,
And from that formless ir-and-rationality,
was brought forth the λογοι σπερματικοι
and mapped into ratio seminales,
the seeds of ratio, of calculation,
translated thence back to the beginning:
In the beginning was the Ratio,
the calculation, the function of functions
the algorithm of Nothing
in limitless transformations,
of the magical stone whence comes
the living waters of loves,
and from thence too a wondrous equation:
Λογος,Ratio, function, algorithm, reason,
    calculation,
synonymous synchronicities,
and "the ratio/calculation was to the god;"
He who has eyes to see, and ears to hear,
Let him see, and let him hear;
Let him be meek, to inherit the earth,
And read it in the original Greek;
λογιζειν, to ratio-nalize and analyze,
ποιειν, to make, *gemacht*, to fashion,
to synthe-poeti-cize the Verbum Dei,
the algorithmic grammar of God.

THE VISION OF PARACELSUS

Analecta Incantationorum 1

Esteemed savant and philosoph,
Venerable alchemic and socratic man:
Was it not heady at first,
this newly found and far-flung freedom of thine?
Distallation of flasks of alchemic fluids bestirred
With mystery and mercury and compounded by
    incantations,
With skeletal bones of vivisected nature its secrets
    to divine,
By annotated bestiarial runes in tomes of the arcane
    secrets
Of metaphysics, and to grasp the dialectical key, to
    pry it open,
And with that intelligible blade divide the
    indivisible,
And all history, man, and nations
By magical science and scientific magick the future
    opine,
And thus progress on occult quest to weigh
    unerringly
All That Has Been and All That Shell Be
To be categorically subsumed in the genus *Omnes*
    quod est..
Breathe the vaporous and sickly sweetness
Of the incense and sip the red tincture and elixir,
Inhale sweet insufflation and drink
And prophecy of Gog and of Magog
And of the Great Illuminator and Beast and
    Mystagogue
And trace by all the thrones and dominations
His broad and easy path to abominations of
    desolations
And all destruction of logick and metaphysick and
    such angelic arts
By demon ideology and other anti-seraphic papal
    muses
With clear and recondite polysyllabilifications;
Let us trace the pentagram of the burgeoning
    logothanatos
At the end of history,
Speak, Stone, Speak
Of atoms smashed and goldenly transubstantiated
Of the division of substance primordial
And grand accidents recombinated
Of gogic dialectical magogic dominion
    reinterpreted

In the towered Babel of magisterium;
Prophecy truly of a false prophecy
Measure its bounds and divine its limits
Speak, Stone, Speak
Of the volume integral and compass
Of false claimants of Yahwism and its ends of the
    world
And of Jahbulon in the tenth degree.

Analecta Incantationorum II

By vats and cauldrons of noble mixture
Is extracted a Stone of subtle poesie and art,
Upon the history of ideas and the history of an idea
Und leider auch Theologie;
Therefore by all the dread powers
Of the terrible necromancy of History
By all the doctrines of the Cosmic Sympathy
By the laws of the universal analogy
And the music of the spheres
Speak, Stone, Speak:
Gather all the facts for a statistical average
And by alchemic sociology, blackest black science
Of mass psychology, predict, nay engineer,
A precise measure of this veil of tears.
Up, ye quadrants and pyramids and trigonometry,
And every tarotic power of predictive science,
And lay the spread that is crossed and broken
And name the cards as they fall,
Solve for "x" and integrate
The Fool, the Card of Death, the Hanged Man,
    and Tower;
Inhale the sickly radiation of frankincense
And sip from the Nine of Cups of our tincture
And prophecy of a false prophecy and its false
    fulfillment
And give the cross product of a false magisterium
Of infallibility in the end of the world
Of a book and Giovanni Mastai-Ferreti who
Being pious too sat upon the seventh step upon the
    seven Hills
In the seventh age.

Analecta Incantationorum III

Triple, triple, triple and ripple;
Stone, Prophecy, and cauldron, cripple;
What is bound, to loose, decant,

And what is loose, to bind, incant:
Speak, Stone, Speak:
Of the halt infirm lameness of invented gods
Of Beelzebub and his drear disobedience.
Speak, of the dread ministrations of Yahweh
Of the spells of agnostics and roasting of heretics
And of the acts of atheistic, sound party men
All grounded equally in faith,
Of logothanatos and of the great malediction
Of all brutality made imperfect in doubt;
Of the feebleness of fire and its not quite ultimate
    unreason,
Of the abysmal vulgarity and the miserific vision;
Speak, Stone, Speak,
Of how was and is and shall be invoked
The ancient drear and shadow weakness
Against the care of love and the comeliness of
    compassion,
Sip and breath and Prophecy
Of the effusive bloody sacrifice
Of seven apostasies from the ancient philosophy
And thrice repeated throes:
The First and Second fulfilled
When first, second and third Charlemagnes
Wrought their pains in the names
Of Moses' bloody Usurper
Von Rhon durch Donau bis Donetz
But the Third and its woes are yet to come.

Pro Circumambulatus Pentagramus

With the finely sharpened point of bluest chalk
Trace the delicate lines of pentacle and pentagram,
Inscribe the pentagon and circle
Imbibe delirious poisons and sweet potions
And red tinctures of youth;
Fetch the androgynous pots and powders
And read the perfumed pages of undateable scrolls,
Pour and stir with the 'phemeral bones
And dream in visions of the aeons and their lofty arts,
Of cured wood and stretched catgut taut,
Of plectra, bows, jacks and keys and frets
And fluted pipes and knobby stops
And of finely-spun lines of golden musick
    composed,
Of sculpting stones for cathedrals and cloisters
And chisel new Davids each ripple and muscle,
Dream of camels' hair and linseed oil,
Of berries' pigments and roots,

And of thick impasto and luminous gesso,
Dream of arcane lost wisdom, of lexicography,
And philology, of careful etymology and turn of
    phrase,
Of artful plots and Dante Alighieri.
Speak, Stone, Speak,
Sip and breathe and Prophecy
Of a trillionfold speechless holocaust,
Of the mute offerings of dumb genuscide,
Of brutish innocents accounted nothing worth
By Theologue and Scientogue alike;
Speak of seven secrets lacking three
When a vast bleakness is all shall be,
And man stands alone upon the Earth.

Analecta Incantationorum IV

There may be some who do not see
The great benefits to Society
Of the sacred craft of Free Alchemy;
It is a most Stoney Science,
A devlishly gluggingly good Science,
Full-laden with elixirish Technologie
For to imbibe and to incant and in full sate
With molecular formula to recombinate.
Trace the ancient pentagram and see
That with it we shall send pictures
Through the *aether lumeniferous* with an extra-
    sensory
Perceptive telemetry.
Prophecy! Prophecy! Prophecy!
And grow salt-water corn with the latest advances
Of our newest branch, the Department of
    Oceanography.
Imbibe, incant, and inhale
The stenchy incense of a heavenly hell
And Speak, Stone, Speak,
And Prophecy of man who hath conquered all
    kingdoms
Round and round the boiling kettle
And stir the global sauce to test his mettle,
Analyze and quantify and destroy the soul
While he weeps for the vanishing egret
Or aborts the foetus without regret
And for the impotent testicle or barren womb
Mourns while preparing
Wombs for hire for a nine month task
Of infants in test tubes born
To be nurtured by an Ehrlenmeyer flask.

Analecta Incantationorum V

Full round the compass
Of the cauldron world
Five times hath been squared
The inflicted lines and points
By crystal stone and pentacle
Bisected.
But of all these Prophecies
And unutterably hideous
Unuttered perversities
Gather the seventh lacking one
Into the Great and Final;
With wretching fear
Stretch forth the wand
To cast off every predilection
And blot out all *a priori*
And limitation of the invocation.
Without Science, and Magickless,
Whisper and dread it,
The preconditionless pleas
For the Prophecy of all Prophecies
And Primal Matter of all materialities;
Speak, Stone, Speak,
Of thine own mind
And know no preset bound
Save the one to be found
The Arcane Key
That unlocks them all
Trace the line and tread it
The presupposition
Of innumerable middle terms
And the conclusion in
The cyborg beastly branding of epiderms
And utter the unknown but manifest
Tautology openly hidden in all
Eschatology.

BIBLIOGRAPHY
of Works Cited or Consulted

Akerley, Ben Edward. *The X-Rated Bible: An Irreverent Survey of Sex in the Scriptures.* Port Townsend, Washington. Feral House. 1998. ISBN 0-922915-55-5.

Albertus Magnus. *Egyptian Secrets: or White and Black Art for Man and Beast.* Kessinger Publising Company. No Date. ISBN 1-5645-9356-8.

Alford, Michelle. "AFIRM Reduces Scare Formation with their Advances in Regenerative Medicine." http://www.scars1. com/news/mainstory.cfm/55.

Alighieri, Dante. *The Divine Comedy: The Inferno, The Purgatorio, and The Paradiso.* Trans. John Ciardi. New York: New American Library. 2003. ISBN 978-0-451-20863-7.

Assmann, Jan. *Moses the Egyptian: The Memory of Egypt in Western Monotheism.* Harvard University Press. 1997. ISBN 0-674-58739-1.

Assmann, Jan. *The Price of Monotheism.* Trans. Robert Savage. Stanford University Press. 2010. ISBN 13: 978-0-8047-6160-4.

Atala, Anthony. "Regenerative Medicine's Promising Future," CNN, July 10, 2011. www.cnn.com/2011/OPINION/07/10/ atala.grow.kidney/index.html

Bacon, Francis (Lord Verulam). *The Advancement of Learning and New Atlantis.* London: Oxford University Press. 1966. No ISBN.

Barnstone, Willis, ed. *The Other Bible.* New York. Harper Collins. 2005. ISBN 978-0-06-081598-1.

Bieri, James. *Percy Bysshe Shelley: A Biography.* John Hopkins University Press. 2008. ISBN 13: 978-0-8018-8861-8.

Booth, Mark. *The Secret History of the World As Laid Down by the Secret Societies.* The Overlook Press. 2008. ISBN 1-59020-031-4.

Campbell, Joseph. *The Hero With a Thousand Faces.* Novato, California: New World Library. 2008. ISBN 978-1-57731-593-1.

Campbell, Joseph. *The Masks of God: Volume I: Primitive Mythology.* New York: Penguin Compass. 1991. ISBN 978-0-14-019443-2.

Campbell, Joseph. *The Masks of God: Volume II: Oriental Mythology.* New York: Penguin Compass. 1991. ISBN 978-0-14-019442-5.

Campbell, Joseph. *The Masks of God: Volume III: Occidental Mythology.* New York: Penguin Compass. 1991. ISBN 978-0-14-019441-8.

Carpenter, Edward. *The Art of Creation.* Bibliolife. No Date. ISBN 978-1-1103-484133.

Carpenter, Edward. *Intermediate Sex.* Kessinger Legacy Reprints. No Date. ISBN 978-1162575476.

Carpenter, Edward. *Intermediate Types Among Primitive Folk: A Study in Social Evolution(1921).* Cornell University Library Digital Collections. No Date. ISBN 918-1112-035944.

Courtis, Jack. "The *Divine Comedy* and Kabala." Internet article. www.crcsite.org/Dante/htm.

Darwin, Erasmus, M.D.F.R.S.. *The Temple of Nature; or, the Origin of Society: A Poem, with Philosophical Notes.* New York: T. And J. Swords. 1804. British Library Historical Collection Reprint. ASIN: B003OBZ6N8

De Nicolás, Antonio, Ph.D.. *Meditations Through the Rg Veda: Four Dimensional Man,* New Edition. New York. Authors Choice Press. 2003. ISBN 0-595-26925-7.

De Santillanam Giorgio, and Von Dechend, Hertha. *Hamlet's Mill: An Essay on Myth and the Framework of Time.* Boston. Gambit Incorporated. 1969. No ISBN.

Dreger, Alice Domurat. *Hermaphrodites and the Medical Invention of Sex.* Harvard University Press. 1998.

Duncan, Malcolm C. *Duncan's Masonic Ritual and Monitor.* Kessinger Legacy Reprints. No Date. ISBN 9781162563527.

Duns Scotus Eriugena. *Periphysion (The Division of Nature).* Trans. I.P. Sheldon-Williams, revised by John O'Meara. Washington, D.C. Dumbarton Oaks. 1987.

Duns Scotus Eriugena, *Periphyseon: On the Division of Nature.* Trans. and ed. Myra L. Uhlfelder. Indianapolis. Boobs-Merril. 1976. ISBN 0-672-60377-2.

Eliade, Mircea. *Mephistopheles and the Androgyne: Studies un Religious Myth and Symbol.* New York. Sheed and Ward. 1965. No ISBN.

Farrell, Joseph P. *Babylon's Banksters: The Alchemy of Deep Physics, High Finance, and Ancient Religion.* Feral House. 2010.

Farrell, Joseph P. *The Cosmic War: Interplanetary Warfare, Modern Physics, and Ancient Texts.* Adventures Unlimited Press. 2007. ISBN 978--1-931882-75-0.

Feldstein, Stephanie. "Human-Animal Hybrids and Other Crimes Against Nature." February 24, 2010. http://news.change.org/stories/human-animal-hybrids-and-other-crimes-against-nature

Fisher, Paul A. *Behind the Lodge Door.* Washington, D.C. Shield Publishing , Inc. 1988. ISBN 0-944700-00-4.

Fox, Maggie. "US company claims cloned humans, made stem cells." http://www.reuters.com/article/2008/01/21/ idUSN1721774620080121.

Garreau, Joel. *Radical Evolution: The Promise and Peril of Enhancing Our Minds, Our Bodies - and What It Means to Be Human.* New York: Broadway Books. 2005. ISBN 978-0-7679-1503-8.

Hall, Manly P. *The Secret Teaching of All Ages: An Encyclopedic Outline of Masonic, Hermetic, Qabbalistic and Rosicrucian Symbolical Philosophy: Reader's Edition.* New York: Jeremy Tarcher/Penguin. 2003. ISBN 1-58542-250-9.

Hutchinson, Martin. "Mixed-sex human embryo created." http://news.bbc.co.uk/2/hi/health/3036458.stm

Iamblichos the Neoplatonist. *The Theology of Arithmetic.* Trans. Robin Waterfield. Grand Rapids, Michigan: Phanes Press. 1988. ISBN 978-0933999725.

Jacobsen, Thorkild. *The Harps that Once...: Sumerian Poetry in Translation.* New Haven: Yale University Press. 1987. ISBN 0-300-07278-3.

Jha, Alok. "First British human-animal hybrid embryos created by scientists." *The Guardian.* Wednesday 2 April 2008. http://www.guardian.co.uk/science/2008/apr/02/medicalresearch.ethicsofscience

Knight, Christopher, and Lomas, Robert. *Uriel's Machine: Uncovering the Secrets of Stonehenge, Noah's Flood, and the Dawn of Civilization.* New York. Barnes and Noble Books. 2004. ISBN 0-7607-5342-3.

Jones, Violet. "Chimeras, Cloning and Freak Animal-Human Hybrids." *Infowars.com.* November 23, 2004. http://www.infowars.com/articles/brave_new_world/chimera.htm

Levi, Eliphas. *The History of Magic.* Trans. A.E. Waite. London: Reider. 1982. ISBN 978-0091500412.

Lomas, Robert. *Freemasonry and the Birth of Modern Science.* Gloucester, Massachusetts. Fair Winds Press. 2004. ISBN 1-59233-064-9.

Magee, Glenn Alexander. *Hegel and the Hermetic Tradition.* Cornell University Press. 2008. ISBN 978-0-8014-7450-7.

Meeks, Wayne A. "The Image of the Androgyne: Some Uses of a Symbol in Earliest Christianity," *History Of Religions,* Vol. 13, No 3 (February 1974), pp. 165-208.

Millegan, Kris, ed. *Fleshing Out Skull and Bones: Investigations into America's Most Powerful Secret Society.* Walterville, Oregon. Trine Day. 2003. ISBN 0-9720207-2-1.

Mott, Maryann. "Animal-Human Hybrids Spark Controversy." *National Geographic News.* January 25, 2005. http://news.nationalgeographic.com/news/2005/01/0125_050125_chimeras_2.html

Narby, Jeremy. *The Cosmic Serpent: DNA and the Origins of Knowledge.* New York: Jeremy Tarcher/Putnam. 1998. ISBN 0-87477-964-2.

Newman, William R. *Promethean Ambitions: Alchemy and the Quest to Perfect Nature.* Chicago. The University of Chicago Press. 2004. ISBN 0-226-57712-0.

Newton, Isaac. *The Principia.* Trans. Andred Motte. *Great Minds Series.* Prometheus Books. 1995. ISBN 0-87975-980-1.

No Author. "Ardhanarisvara: Androgynous Form of Siva." http://www.bhagavadgitausa.com.cnchost.com/HE-SHE-ANDROGYNY-NATARAJA.htm

No Author. "Parahuman." *Wikipedia.* http://en.wikipedia.org /wiki/Parahuman

Pater, Walter. *The Renaissance: Studies in Art and Poetry.* Mineola, New York. Dover Publications, Inc. 2005. ISBN 978-0-486-44025-5.

Peckhaus, Volker. "Calculus Raiocinator vs. Characteristica Universalis? The Two Traditions in Logic Revisited."

Picknett, Lynn, and Prince, Clive. *The Forbidden Universe: the Occult Origins of Science and the Search for the Mind of God.* Skyhorse Publishing. 2011. ISBN 10-1-61608-028-0.

Prabhavananda, Swami, and Isherwood, Christopher, trans. *Bhagavad-Gita: The Song of God.* New York: Signet (Penguin). 2002. ISBN 978-0-451-52844-5.

Regal. Philip J. "Metaphysics in Genetic Engineering: Cryptic Philosophy and Ideology in the 'Science' of Risk Assessment. *Coping with Deliberate Release: The Limits of Risk Assessment*, ed. Ad Van Dommelen. Tilburg/Buenos Aires: International Centre for Human and Public Affairs. pp. 15-32. This paper may also be found at http://www.psrast.org/pjrbiosafety.htm

Robin, Marie-Monique. *The World According to Monsanto: Pollution, Corruption, and the Control of Our Food Supply: An Investigation into the World's Most Controversial Company.* New York: The New Press. 2010. ISBN 978-1-59558-426-7.

Rudolph, Kurt. "Ein Grundtyp gnostischer Urmensch-Aram-Spekulation,"1-20, *Zeitschrift für Religions- und Geistesgeschichte.* 1957. IX Jahrgang. Heft 1.

Schwarz, Arturo. "Alchemy, Androgyny and Visual Artists," *Leonardo,* VOl. 13, pp. 57-62. Pergamos Press. 1980. www.jstor.org/pss/1577928.

Scott, Walter, trans. and ed. *Hermetica: The Ancient Greek and Latin Writings Which Contain Religious or Philosophic Teachings Ascribed to Hermes Trismegistus.* Kessinger Publishing Company. No Date. ISBN 1-56459-481-5.

Shreeve, James. *The Genome War: How Craig Venter Tried to Capture the Code of Life and Save the World.* New York: Ballantine Books. 2004. ISBN 978-0-345-43374-9.

Sutton, Anthony C. *America's Secret Establishment: An Introduction to the Order of Skull and Bones.* Liberty House Press. 1986. ISBN 0-937765-02-3.

Sykes, Bryan. *Adam's Curse: The Science that Reveals Our Genetic Destiny.* New York: W.W. Norton. 2004. ISBN 0-393-32680-2.

Tarpley, Webster. *Against Oligarchy.* Online book. http://tarpley.net/online-books/against-oligarchy/giammaria-ortes-the-decadent-venetian-kook-who-originated-the-myth-of-carrying-capacity/

von Franz, Marie-Luise, Ph.D., ed., and Hull, R.F.C. and Glover, A.S.B., trans. *Aurora Consurgens: A Document Attributed to Thomas Aquinas on the Problem of Opposites in Alchemy.* Toronto. Inner City Books. 2000. ISBN 978-0-919123-90-8.

Weidner, Jay, and Bridges, Vincent. *The Mysteries of the Great Cross of Hendaye: Alchemy and the End of Time.* Rochester, Vermont. Destiny Books. 2003. ISBN 089281084-X.

Wilde, Oscar. *The Complete Works of Oscar Wilde.* Intr. Merlin Holland. HarperCollins Publishers. 2003. ISBN 978-0007144358.

Wilde, Oscar. *The Picture of Dorian Gray.* Intro. Camaille Canti. Barnes and Noble Classics. 2003. ISBN 978-1-59308-025-9.

Yates, Frances, A. *The Art of Memory.* Volume III, *Selected Works of Frances Yates.* London: Routledge. 2010. ISBN 978-0-415-60605-9.

Yates, Frances A. *Giordano Bruno and the Hermetic Tradition.* Vol II, *Frances Yates: Selected Works.* London: Routledge. 1964. No ISBN.

FERALHOUSE.COM